'The nuclear industry invites us, all the time, to look forward
compelling study shows why: its legacy, all around the world, ..., ...
term solutions to the problem of nuclear waste in sight, and countless communities blighted, in
one way or another, by the nuclear incubus in their midst.'

Jonathon Porritt, Founder Director, Forum for the Future, UK

'Andrew Blowers provides the finest single volume on nuclear power I have encountered in my half
century of energy research. Blowers' seminal publication is a masterwork built atop a contemplative
foundation of interdisciplinary reading, innumerable interviews and active international fieldwork.
This, his newest book, provides us with a clear-eyed look inside the periphery of the most
controversial source of electricity in the world. All those considering the path we should take toward
the energy future our descendants will inherit need first to absorb what Andrew Blowers has to say.'

*Martin J. Pasqualetti, Professor of Geographical Sciences and
Urban Planning, Arizona State University, USA
Winner of the Alexander and Ilse Melamid gold medal from the American Geographical Society*

'Andy Blowers knows the communities around nuclear sites, across many countries, better than
anyone. His book is utterly compelling, beautifully written and explains how a variety of
consecutive discourses has influenced the evolution of these "peripheral" communities, giving
them the ambiguous status of being both marginal and dependent, but also endowed with political
influence. In this process, he shows with great clarity and insight why it is so difficult to move
forward with long-term solutions for the nuclear waste problem and why the waste is so unlikely
to move from its current locations.'

*Gordon MacKerron, Professor of Science and Technology Policy, SPRU, University of Sussex, UK
Chair of Committee on Radioactive Waste Management, 2003–2007*

'The carefully distilled fruits of a long career researching social circumstances of communities around
"nuclear oases" worldwide, this book is topical, readable, authoritative – and sometimes deeply
moving. The message is of crucial political importance in all countries where these challenges persist.'

Andy Stirling, the Science Policy Research Unit at the University of Sussex, UK

'Andrew Blowers has been involved in this subject since the early 1980s when, as a local councillor
in England, he was confronted by an industry attempting to dispose of radioactive waste on his
doorstep. This is not a dry academic treatise. Through numerous meetings with others affected
by the industry's vain attempts to solve its intractable disposal problem he has been able to
empathetically document grassroots community responses to this issue.'

Pete Roche, Editor of no2nuclearpower.org.uk

'Where should nuclear waste go? This carefully researched book explores the political and ethical
issues by identifying key characteristics for sites that have been selected or proposed. They have
almost all been in peripheral, often economically weak areas, usually with existing nuclear
projects, where local resistance and opposition was unlikely or muted. A grim legacy.'

David Elliott, Emeritus Professor of Technology Policy, the Open University, UK

'Forty years ago, Brian Flowers in his seminal report for the Royal Commission on Environmental
Pollution identified the lack of proven solutions to radioactive waste disposal as the Achilles Heel
of the nuclear industry. Andrew Blowers' perceptive and well-written account of attempts at
sites in four key countries, UK, USA, France and Germany clearly demonstrates why it is
proving so difficult to overcome this issue. It should be essential reading for anyone with an
interest in this issue.'

Stephen Thomas, Emeritus Professor of Energy Policy, University of Greenwich, UK

'Andy Blowers gives us an expertly researched and insightfully written book that explores the interplay of pivotal social, political, and economic relationships functioning inside four nuclear communities facing the multi-generational management and clean-up of radioactive waste. A must read book, not only for the public but also decision makers facing radioactive waste issues.'

Roy Gephart, Former Chief Environmental Scientist from the Pacific Northwest National Laboratory, Washington State, USA

'Both an incisive and searing indictment of the short-sightedness of the nuclear industry and a penetrating analysis of how the industry both generates and perpetuates peripheral communities, which shoulder enormous risks reprocessing and burying nuclear waste. Andrew Blowers writes with the full authority expected of a world leading nuclear expert of over 30 years' standing. His work is burnished with a deep humanism that keeps one eye firmly on the stakeholders that are usually neglected by policy makers; the generations of the future.'

David Humphreys, Professor of Environmental Policy, the Open University, UK

'*The Legacy of Nuclear Power* makes for a passionate, well argued, and articulate read on many of the social, economic, and political costs to nuclear power. It pushes the boundaries not only for forcing us to critically assess and perhaps rethink contemporary nuclear and energy policy, but also because it brings to the forefront disturbing questions related to social justice, equity, and the environment. Important reading for anybody seriously interested in the nuclear renaissance.'

Benjamin K. Sovacool, author of The National Politics of Nuclear Power, *Aarhus University, Denmark*

'The book tells a fascinating yet disconcerting backstory of the "participatory turn" in radioactive waste management. Through his in-depth four-country comparison, Blowers provides a compelling analysis of the power play behind the systematic tendency of peripheral communities ending up as the preferred hosts for nuclear waste disposal sites. Armed with his unique first-hand experience as activist, researcher, policy advisor, and policymaker, Blowers invites the reader to the backstage of policymaking on "the waste problem" – the enduring the Achilles' heel of the global nuclear industry.'

Markku Lehtonen, Research Fellow, Ecole des Hautes Etudes en Sciences Sociale, France

'Andrew Blowers offers an excellent insight into the geography of nuclear activities, describing how global nuclear dangers are connected to local sites where ordinary people live their everyday lives. There is a lot to be learned from this detailed and engaging story of nuclear sites in four separate countries.'

Göran Sundqvist, Professor of Science and Technology Studies, University of Gothenburg, Sweden

'This book makes a great case for remote and culturally distinct communities, such as my home region Wendland. I could not agree more with Blowers' analysis of the Gorleben protests claiming their success was due to them being "at once both flexible and obstinate, resisting change while also ultimately embracing it". In contributing to the development of a responsible and secure management concept for nuclear waste, this small protest community made big achievements in dealing with the nuclear legacy of our society – over decades, to this day. This is indeed zeitgeist and provides clear evidence for the principle which should be applied here: care before rush.'

Rebecca Harms, President of the Greens/EFA European Parliament, Germany

THE LEGACY OF NUCLEAR POWER

Nuclear energy leaves behind an infinitely dangerous legacy of radioactive wastes in places that are remote and polluted landscapes of risk. Four of these places – Hanford (USA), where the plutonium for the first atomic bombs was made; Sellafield, where the UK's nuclear legacy is concentrated and controversial; La Hague, the heart of the French nuclear industry; and Gorleben, the focal point of nuclear resistance in Germany – provide the narratives for this unique account of the legacy of nuclear power.

The Legacy of Nuclear Power takes a historical and geographical perspective going back to the origins of these places and the ever-changing relationship between local communities and the nuclear industry. The case studies are based on a variety of academic and policy sources and on conversations with a vast array of people over many years. Each story is mediated through an original theoretical framework focused on the concept of 'peripheral communities' developing through changing discourses of nuclear energy. This interdisciplinary book brings together social, political and ethical themes to produce a work that tells not just a story but also provides profound insights into how the nuclear legacy should be managed in the future.

The book is designed to be enjoyed by academics, policy makers and professionals interested in energy, environmental planning and politics, and by a wider group of stakeholders and the public concerned about our nuclear legacy.

Andrew Blowers OBE is Emeritus Professor of Social Sciences at the Open University. Over a long career he has been involved in the field of environmental politics and policy making as an academic, politician, government adviser, nuclear company director and prominent environmental activist. As a member of the first Committee on Radioactive Waste Management he was directly responsible for some of the UK's policy on legacy waste. *The Legacy of Nuclear Power* brings together his varied experience and expertise and reflects his lifetime concern with the fate of nuclear communities now and in the future.

THE LEGACY OF NUCLEAR POWER

Andrew Blowers

Routledge
Taylor & Francis Group

LONDON AND NEW YORK

from Routledge

First published 2017
by Routledge
2 Park Square, Milton Park, Abingdon, Oxon OX14 4RN

and by Routledge
711 Third Avenue, New York, NY 10017

Routledge is an imprint of the Taylor & Francis Group, an informa business

British Library Cataloguing-in-Publication Data
A catalogue record for this book is available from the British Library

Library of Congress Cataloging-in-Publication Data
Names: Blowers, Andrew.Title: The legacy of nuclear power / Andrew Blowers.
Description: Abingdon, Oxon ; New York, NY : Routledge, 2016. |
Includes bibliographical references.
Identifiers: LCCN 2016009928| ISBN 9780415870009 (hb) |
ISBN 9780415869997 (pb) | ISBN 9781315770048 (ebook)
Subjects: LCSH: Nuclear energy--Risk assessment. | Nuclear energy--Safety measures. | Nuclear power plants--Risk assessment. | Nuclear power plants--Safety measures. | Nuclear power plants--Environmental aspects. | Nuclear energy--History. | Radioactive wastes--Storage.
Classification: LCC TK9152.16 .B56 2016 | DDC 363.17/99--dc23LC
record available at http://lccn.loc.gov/2016009928

ISBN: 978-0-415-87000-9 (hbk)
ISBN: 978-0-415-86999-7 (pbk)
ISBN: 978-1-315-77004-8 (ebk)

Typeset in Bembo
by Saxon Graphics Ltd, Derby

For Varrie and our children and grandchildren in the hope of a better legacy

CONTENTS

ILLUSTRATIONS

Figures

Table

PREFACE

One Sunday evening, in late October 1983, I was sitting in my study in Bedford when the phone rang. It was a reporter from the *Sunday Times* who asked me if I had seen that morning's edition which ran a story about a search, by a newly created body called Nirex,[1] for a possible site for a nuclear waste dump in Bedfordshire. At the time I was a county councillor with a particular interest in environmental policy and planning. I had just completed a book based on my participant observation of a conflict over the development of new super brickworks in Bedfordshire's Marston Vale. On the one side was the London Brick Company, whose operations used the self-firing properties of the region's Oxford Clay to manufacture vast quantities of bricks in the brickyards of the Vale. The brickmaking process, backed by business interests and trade unions, brought jobs and wealth and had attracted successive waves of immigrants to the area but it also scarred the landscape and created emissions of pollutants, sulphur dioxide, fluorides and mercaptans. On the other side was an alliance of landowners, farmers and environmentalists who perceived an opportunity of wresting environmental benefits in return for allowing the new superworks to be built. There were many twists and turns and unanticipated shifts in power relations as well as unintended outcomes before the conflict reached its dramatic conclusion. But that is another story which I have already told in my book, *Something in the Air* (1984).

Anyway, the point of this seeming digression is to reveal the autobiographical context in which I first encountered what was to become a consuming, indeed passionate interest, if that is not too strong a term, in the social implications and consequences of radioactive waste. On that October evening I was already disposed to continuing a political and academic commitment to environmental conflicts. It suited my long-established geographical and historical interest in landscape, places and communities as well as my ideological concerns with democracy and equality and their spatial unevenness and intergenerational variation. Moreover, I was particularly interested in understanding the local context of conflict, so, when the reporter indicated that the area of search for a site for a nuclear dump was in an area about 50 miles north of London and probably in Bedfordshire, I was immediately intrigued. Asked if I had any idea where such a site might be I replied, 'No, but I'll endeavour to find out'.

My instant, and indeed considered, reaction was that the site must be one of the abandoned brick pits, extensive, exhumed voids scattered across the Marston Vale. Using serendipitous, if primitive, site selection criteria I assumed a site for a nuclear waste dump would need to be in friendly ownership, have no competing land use claims and would preferably be in an area of dereliction with little pretence to environmental value. The brick pits amply fitted my expectations and I spent the following day in a fruitless search. Later that afternoon I received another phone call, this time from the BBC, informing me that the site had been found and would I come immediately to London to be interviewed live on the six o'clock TV news. The actual site was just south of Bedford, on a former wartime munitions dump, near the village of Elstow, birthplace of one of Bedford's most famous sons, John Bunyan, the radical 17th century preacher, whose renowned allegory, *The Pilgrim's Progress,* charts Christian's pilgrimage through a number of recognisable Bedfordshire landmarks. One of these is the Slough of Despond into which Christian tumbles shortly after setting forth. This provided an apt metaphor for my TV appearance when I announced to the nation that John Bunyan would be turning in his grave at the thought that the Slough of Despond was to reappear in contemporary form as a nuclear waste dump. I declared that the people of Bedfordshire would fiercely oppose such a desecration of their environment. My instinctive belief that anything 'nuclear' would be resisted in a greenfield location proved correct for, within days, packed public meetings, extensive media coverage and local political and environmental action began the long process of opposition (also described in a 1991 book *The International Politics of Nuclear Waste*) which, in this case, ended with the withdrawal of the Nirex project four years later.

So began a political and intellectual odyssey that was to last the rest of my life. My political journey, beginning at the level of local politics, soon developed at national level through my role on government advisory bodies, where I was involved in policy development, bringing a social and ethical context to what had hitherto been a largely technical and scientific matter. The opening up to public participation and engagement was the key element in the transformation from a closed and secretive to an open and transparent policy making process. My journey literally took me to many places, visiting nuclear facilities often several times, surface and underground, gathering information and experience of radioactive management overseas. Although I was fascinated by the technology, scale and dangers associated with managing nuclear waste, the core of my interest was the people and places constituting the communities that are the guardians of our nuclear legacy. Their story is the inspiration for this book.

At the intellectual level I became fascinated with the idea of the nuclear industry as the epitome of 'Risk Society' yet, at the same time, an industry that sought to cultivate an image of safety, security and sustainability. There seemed to me something preposterous about this juxtaposition of danger and desire. During the years I was writing this book the contradictions were amplified. The idea of the Nuclear Renaissance with its overtones of innovation, reaffirmation and progress was countered by the vision of Nuclear Nightmare, silently present at nuclear sites, but in reality visited on Japan with the Fukushima disaster in 2011. The choices seemed stark: on the one hand, continuing development of nuclear technology as a sustainable way forward for energy policy; on the other, abandonment of nuclear power as too costly, too dangerous and too contaminating to be a suitable energy form for the future. This constant oscillation between nuclear or not continues to play itself out in different ways, in different social and political conditions in different countries. These variations in the perennial contest are present in the communities and countries featured in this book, from the resigned embrace of nuclear clean-up in Hanford USA, to the resolute approach towards geological disposal as a permanent solution pursued in France, to the conflicting priorities that have beset Sellafield in the UK and, finally, the vigorous resistance to nuclear developments that characterises Gorleben in Germany. Although each is very different they have each been socially shaped by the evolving power relations of the nuclear industry and its antagonists. They each evince the processes and character of what may be called 'peripheral communities'.

The idea of peripheral communities and the processes of 'peripheralisation' (unlovely word) through which hazardous, including nuclear, activities are attracted to or repelled from particular communities has its origins in what my colleague Pieter Leroy and I dubbed the 'Nijmegen Summit'. On the terrace in his garden we tossed ideas back and forth which eventually took shape in the thesis of peripheralisation which we later set out in a paper which is discussed in the opening chapter of this book. In the years that have elapsed since, I have found this concept a useful explanatory device in analysing the power relationships at work within nuclear communities and the processes of inequality that underlie them. As both politician and academic I have been fascinated throughout my life with the meaning and purpose of power. It is to another colleague, Steve Hinchliffe, with whom I produced an Open University course on *Environment* that I owe the interpretation of power that provides the dynamic analysis of the relations between industry and community that is the theme of this book. Together we clarified a distinction between power as resource and power as discourse. While discourses of power (set out in terms of nuclear discourses in the book) changing and sometimes competing over time establish the context and boundaries within which conflicts occur, outcomes will relate to the resources (material, people, information) that can be mobilised and deployed at specific times and places.

My interest in nuclear waste thus stimulated by political involvement and academic curiosity has always had a more practical dimension, in policy making. As a councillor I was able to influence directly the course of a specific conflict and as an academic I was able to explore ideas and explanations for processes and patterns of nuclear development but I also became involved in policy development, as government adviser on the Radioactive Waste Management Advisory Committee and then on the first Committee on Radioactive Waste Management (CoRWM). With CoRWM I helped to establish the social, ethical and political context of policy making and drafted its basic set of recommendations (described in Chapter 3) to government. At the same time I led a group concerned with the implementation of policy, based on principles of voluntarism, participation and partnership, in the search for a suitable site for a geological disposal facility for the nation's accumulation of legacy wastes.

Throughout my political career in nuclear policy making I remained anti-nuclear but willingly cooperative in helping to create a consensual approach to managing nuclear wastes based on the mantras of openness, transparency and participation. I was even invited to join the Board of Nirex, my former adversary, then anxious to transform its reputation in the wake of its failures to find sites for nuclear waste culminating in the defeat of its proposal for a deep underground laboratory at Sellafield in 1997. My appointment was, I think, seen as part of an effort to sanitise the company and to give it a softer, more community-friendly image. It was to no avail since Nirex was later absorbed into the Nuclear Decommissioning Authority. By that time, in the early years of this century, the policy consensus was breaking down as new nuclear rose Phoenix-like, promoted as the putative answer to 'keeping the lights on', 'bridging the energy gap' and 'saving the planet from climate change'. CoRWM had reached its own consensus on radioactive waste management expressed in its key recommendation: 'Within the present state of knowledge, CoRWM considers geological disposal to be the best available approach' for managing higher active nuclear wastes. But, this was predicated on 'being able to sustain the favourable political environment in which these recommendations have been developed'. CoRWM's approach related to legacy wastes, not to wastes from new build which raised quite different political and ethical issues, not least the essentially unknowable timescales for implementation. Adding new build wastes to the existing burden of legacy wastes makes finding a suitable and acceptable site for the disposal of radioactive wastes more difficult.

With the advent of new nuclear build I reverted to a more active anti-nuclear role. My early experience was opposing Nirex as an elected politician; now, after years of involvement in nuclear policy making, I became an activist in opposition to proposals for siting new nuclear power stations and spent fuel stores at various coastal sites around England and Wales. I set up a citizens' action group called Blackwater Against New Nuclear Group (BANNG) specifically to campaign against a new nuclear power station and waste site on a former nuclear site at Bradwell on the Blackwater estuary in Essex, some 50 miles from London. I became co-chair of a national forum which brings together national government and anti-nuclear NGOs. I am an active participant in the complex web of power relations including national governments (British, French, Chinese), international nuclear organisations, nuclear companies, local authorities, regulatory bodies, conservation organisations, environmental groups and the general public who each have resources to deploy and a part to play in determining the fate of the nuclear industry and the communities in which it operates. The course of the conflict is ever shifting and the outcome remains uncertain. Fundamentally, it is not simply about the balance of environmental risks and economic benefits for the present but about what kind of legacy is passed on to generations in the far future.

This book has drawn deeply on my experience of the places I visited and the people I met. The conversations have animated the narratives in the chapters on each country and I have recorded the names of those I met. I am particularly indebted to the help and insights I gained from Roy Gephart for the Hanford chapter, from Stephen Thomas on Sellafield, Markku Lehtonen and Luis Aparicio on France and Peter Ward on Gorleben. John Mathieson of Nirex (now RWML) and Mark Dutton a colleague on CoRWM helped pave my way and occasionally accompanied me on various visits. At a crucial point in my writing my colleague at the OU Petr Jehlicka gave the critical help and encouragement I needed to bring the project to fruition. I must also commend the magnificent contribution of maps and illustrations of another OU colleague, John Hunt. There are many others whose encouragement and support have continued to sustain my interest and commitment especially Mike Pasqualetti, David Lowry, Dave Humphreys, David Potter and Pete Wilkinson all whose friendship I have enjoyed over the years. And there are the myriad of colleagues with whom I worked on committees, political allies and opponents, nuclear industry and regulators, academics and activists who have helped me to clarify and confirm my perspectives.

And, lastly, I wish to pay tribute to my late wife Gill with whom I began this journey and above all I wish to thank Varrie who has given the unconditional emotional, critical and practical support over two decades in the life of this project. Together we have put our energies to practical purpose in the hope that our generation can bequeath a better legacy for the generations that come after than the one we have presently inherited.

Bedford, England, April 2016

Note

1   Nuclear Industry Radioactive Waste Executive, government body established to find suitable sites for the disposal of nuclear wastes.

References

Blowers, A. (1984) *Something in the Air: Corporate Power and the Environment,* London, Harper and Row.

Blowers, A., Lowry, D. and Solomon, B. (1991) *The International Politics of Nuclear Waste,* London, Macmillan.

1

NUCLEAR OASES

The persistence of the periphery

The enduring geography of nuclear waste

The legacy of nuclear power is found in the sites around the world where uranium has been mined, where nuclear power has been produced, where nuclear weapons have been manufactured and tested and where reprocessing of nuclear fuel for plutonium has taken place. The legacy consists in the buildings that once housed reactors, in the pools and storage areas that hold radioactive wastes and in the areas of contaminated land and polluted waters caused by emissions and discharges, leakages and precipitation. The legacy of radioactive waste in its various forms is inevitable and long-lasting arising from routine operations but also from accidents, some of them releasing large quantities of radioactivity over wide areas, in the worst cases causing the evacuation of surrounding populations and perpetual restrictions on human habitation. While Fukushima and Chernobyl are the most notorious nuclear accidents, there have been many others, often unpublicised and sometimes unrevealed until long after they occurred. But, around every nuclear installation there is restricted access for safety and security reasons, and beyond, often an area where various restrictions and emergency planning procedures are in place. The calculation of risk to human health and the environment from exposure to radioactivity is a contested area; all that can really be said is that a risk, real and perceived, exists and persists over time though the nature and extent of the risk is a matter of seemingly irreconcilable debate. The communities living in the shadow of nuclear facilities are, thereby, communities living with risk and it is these communities, the problems that they face and the implications for how we manage the nuclear legacy which are the subject of this book.

The idea that nuclear power station sites or the nuclear reprocessing complexes scattered around the world were destined to survive in perpetuity as nuclear waste dumps was far from the thoughts of those planning, promoting and developing the nuclear economy. Yet, the production of electricity or plutonium will seem only a short phase in the lifetime of nuclear facilities and communities, assuming, of course, that the catastrophic potential of nuclear fission does not, in the meantime, bring about its own demise and that of everything else. For there always looms the possibility, small but there all the same, of a major nuclear accident on the scale of Chernobyl or Fukushima or, more terrible still, the prospect of a nuclear war that escalates out of control (Perrow, 1999; Schlosser, 2013). In the apocalyptic words of the late Ulrich Beck there is always the prospect of instant annihilation: 'But the effect only exists when it occurs, and when it occurs, it no longer exists, because nothing exists any more' (1992, p.38). Nuclear risk is a paradox in time – it is, at once, both everlasting and instantaneous.

For the present and extending into the future, all those places with nuclear facilities possess a long-term and default function which is to become sites for the storage of radioactive wastes. Although almost every nuclear country has, in principle, plans for the eventual burial of the most radioactive of these wastes, only a handful have made any progress towards that end and nowhere in the world is there yet an operating and authorised site for the deep geological disposal of the highest level radioactive wastes.[1] Finding a site that will safely contain the wastes for hundreds of thousands of years and that is also acceptable to society has proved an intractable problem in many countries. Even if acceptable sites are found it will be many decades before wastes can be satisfactorily buried. Focusing attention on geological disposal as the Holy Grail suggests that a permanent solution has already been found to the problem of the nuclear industry's most dangerous legacy. It clears a hurdle that otherwise might impede the industry's further progress and expansion. It leaps ahead to a promised land that may never exist, diverting attention from the problem of the existing legacy, the accumulation of wastes in varying conditions of safety and security that creates landscapes of risk. Whatever the ultimate solution, it is the management of these wastes here and now and extending into the far future that is both the problem and the solution.

In many countries, nuclear power stations are the interim storage sites for the legacy of nuclear wastes pending the eventual arrival of a deep geological repository at some distant time. In a few countries, the UK, France, Russia and the US among them, radioactive wastes have also accumulated at the sites of large reprocessing facilities originally developed for the production of plutonium for nuclear weapons. Reprocessing is the chemical separation of fissionable plutonium from irradiated (or spent) fuel. Reprocessing requires a range of nuclear facilities (reactors, reprocessing plants and waste management facilities) to process and store the complex waste products including highly active liquid wastes, that result. Reprocessing has also been developed for civil purposes to recycle and reuse plutonium in the manufacture of mixed oxide fuel for use in thermal reactors. There is also the prospect, optimistically presented by the nuclear industry, that plutonium may find a future use in fast breeder reactors which have for decades been at an experimental stage of development. However, as the need for plutonium for military purposes has receded with the ending of the Cold War, as markets for reused nuclear fuel have failed to develop and as fast breeder programmes have faltered, so reprocessing has diminished and there has been a palpable shift from production of plutonium to cleaning up and safely managing the dangerous legacy of wastes that has been left. It is, therefore, necessary to place plutonium and other high level waste materials in long-term secure storage until deep geological disposal becomes available. Until then these materials must be kept in high security stores to protect them from diversion and thereby prevent the possibilities of nuclear proliferation. This, then, is the background for my study.

The overarching theme for this book is the relationship between nuclear communities and radioactive waste. In it I intend to illuminate and reflect on three broad themes. First, is the empirical feature of nuclear's legacy: its tendency to be confined to established locations and the implications both for the nuclear industry and the local community. Second, is a conceptual characteristic: the tendency for these locations to manifest what I shall call 'peripheral' characteristics, places that are at the margins geographically and economically, and politically and socially dependent in certain aspects. And third, a more theoretical theme, is a concern with the power relations between industry and community in terms of the discourses they manifest and the resources they are able to deploy. These three themes – empirical, conceptual and theoretical – help to define, understand and explain the relationship between community and legacy and suggest some implications for radioactive waste management strategy which I shall consider in the final chapter.

Communities on a continuum

As to the first theme, the location of the nuclear legacy, there are two abiding characteristics: its persistence both in place and time. The geography of the industry, at least throughout the Western countries with which I am concerned in this book, was established long ago and, with few exceptions, the pattern has remained frozen ever since. Moreover, nuclear activities have extraordinary durability over time, with radioactivity present as nuclear waste extending into the far future, with the possibility of further perpetuation through the construction of new power stations and radioactive waste facilities. The anchoring of the nuclear industry in its existing locations has bequeathed a set of 'nuclear oases', those places where nuclear facilities in some form already exist and therefore where nuclear's legacy will prevail (Blowers et al., 1991; Blowers, 1999). The most extensive nuclear oases with the biggest communities are the major reprocessing sites, including the military nuclear reservations in the United States and Russia (Bradley, 1998; Brown, 2013) and the civil and military complexes in the UK and France. Here, substantial nuclear communities have grown up to support the myriad of complex activities that take place there. These places I have called 'nuclear communities' since their work and lives are focused on managing and cleaning up the legacy of nuclear power and reprocessing. I have chosen four such communities in four different countries as the context for my analysis of the legacy of nuclear power.

Over many years I have visited over 40 nuclear sites in twelve countries. I have toured nuclear power stations, examined nuclear waste facilities and been deep underground to experience research laboratories for geological disposal (in Sweden, France, Germany and Switzerland) and have seen the first (and, so far, the only) operational underground repository in New Mexico, United States. Of all the places I have been there were four to which I became repeatedly drawn and which I decided to study. Each is a major site for managing the nuclear legacy in its country. It is possible to place each site on a continuum based on how long they have been established and how stable the relationship is between nuclear communities and the nuclear industry they support and on which they depend. At the same time they are places at different stages on a continuum of interdependent and overlapping economic, social and political characteristics which together constitute each local community in relation to the nuclear legacy it is hosting. Each of the communities I have chosen is featured in the chapters that follow.

Hanford in Washington state, USA, was established during the Second World War as the site for the production of plutonium which was first used in the atom bomb that destroyed Nagasaki in 1945 (Gerber, 1997; Hevly and Findlay, 1998; Brown, 2013). During the four decades after the War, Hanford expanded its role as the plutonium production site for the arms race and acquired a range of other nuclear activities. Over this period large volumes of highly active nuclear wastes accumulated along with other wastes, leaving a legacy that has made the site one of the most contaminated areas on earth. Since the end of the Cold War, Hanford's production function has wound down and the site is now dedicated fundamentally to decommissioning its reactors and reprocessing plants, managing its wastes and cleaning up the site, a process that will take decades to come. The Hanford site sprawls over 586 square miles in the semi-desert on a bend in the Columbia River. To the south of it, it has developed a substantial urbanised area comprising the Tri-Cities. This, or part of it at least, is an established nuclear community occupying a relatively stable position at one end of the continuum. The nuclear industry may have lost some of its dominance but remains the single most important determinant of the community's identity and culture. And, though separate and still relatively secretive, the huge military Hanford reservation still casts its awesome physical presence over this remote area of America's North West.

Sellafield in Cumbria in the northwest of England lies a little further along my continuum. Established just after the Second World War, like Hanford it was originally set up as a military site for the manufacture of plutonium for the UK's independent nuclear deterrent. Over the years it developed a whole range of non-military nuclear functions including the production of electricity, development of experimental fast reactors and manufacture of mixed oxide fuels. By contrast with Hanford, Sellafield is a very compact site covering less than two square miles and it combines both military and civil functions though, with the Windcale reactors shut down, the military presence is nowadays mainly legacy functions of storage and waste management. Sellafield is at an earlier transitional stage than Hanford in the shift from production to clean-up. Reprocessing, once the lifeblood of the plant dealing with both UK and foreign spent fuel, now mainly a method of waste management rather than plutonium production, is being phased out for technical and commercial reasons. Like Hanford, Sellafield is becoming more and more dedicated to waste management and clean-up of what is commonly regarded as the most dangerous site in Western Europe. The neighbouring area of West Cumbria comprises small towns and villages on the edge of England's celebrated Lake District. Although reprocessing is dying, other possibilities of breathing new life into Sellafield's nuclear complex are tantalising if unlikely such as a new mixed oxide fuel plant or the development of new reactors adjacent to the site. The prospect of the UK's deep geological repository being constructed here, near where most of the country's wastes are located, has twice been rejected but could yet be revived. Unlike Hanford, Sellafield has not achieved a fully settled relationship between industry and community. So, Sellafield still has the appurtenances of a nuclear mirage while the reality is a nuclear oasis in transition to a future of cleaning up the nation's nuclear legacy.

In the case of France, I have chosen two sites for my study which, taken together, are the main places for the management of the country's nuclear legacy. La Hague at the tip of the Cotentin peninsula in Normandy is the location of France's civil nuclear reprocessing industry (reprocessing for military purposes was located at the country's first reprocessing plant at Marcoule in the south of the country). France has a high dependency on nuclear electricity and total production is second only to that in the United States. Moreover, the French industry has, from the outset, adopted the reprocessing cycle so that practically all the spent fuel from reactors across the country is transferred to La Hague, to be reprocessed for plutonium with the resulting high level waste streams converted into vitrified blocks for storage and ultimately disposal. La Hague was first developed during the 1970s and expanded eventually to incorporate two large reprocessing works to deal with both French and overseas spent fuel from foreign customers. The return of wastes to these customers, though contractually necessary, has caused conflict with anti-nuclear groups in Japan and, as we shall see, has been a major source of conflict over nuclear power in Germany. The ultimate destination for French high level waste is very likely to be Bure, in eastern France, the second site of my study. Its present status is as the site for an underground research laboratory which has been excavated. Although not finally confirmed Bure is, to all intents and purposes, the putative site of the national nuclear repository, the final link in the chain from power stations through La Hague to Bure. La Hague, unlike Hanford and to a lesser extent Sellafield, is still a production site but it has a relatively settled relationship with the rural and urban communities that populate this remote corner of France. As yet, there is very little nuclear related development at Bure, a nuclear oasis in prospect though not yet in being. The site is in a thinly populated rural area where the nuclear industry gained a foothold almost by stealth and with very little conflict.

By contrast, the fourth of my case studies, Gorleben in Germany, has been riven with conflict right from its beginnings in the late 1970s. Originally, the site in Lower Saxony near the Elbe River on the border with East Germany was geographically peripheral although since reunification it is more centrally located within the country. Nevertheless, it still has a sense of remoteness and a distinct cultural identity. Gorleben was initially proclaimed as the site for an *Integrierte Entsorgungskonzept* (Integrated Waste Management Concept), an all-singing, all-dancing nuclear complex including an interim radioactive waste store and reprocessing facility associated with a prospective deep disposal repository buried in a salt dome beneath. From the start the project proved deeply controversial and, in the wake of the Three Mile Island disaster in 1979, the plans for a reprocessing plant were dropped. Indeed, it proved politically impossible to find an acceptable site for reprocessing anywhere else, hence Germany came to rely on La Hague and Sellafield with the vitrified wastes arising destined to be returned to the Gorleben interim fuel store. But, the salt dome and the interim store became the focus of anti-nuclear protest. The Gorleben protests had a strong, persistent and implacable local basis but drew in support from across the country in cross-cutting coalitions capable of mounting mass protests. Although the salt dome was excavated as a research facility, it remained controversial and, in the post-Fukushima anti-nuclear triumph that brought about the phase-out of nuclear energy, the salt dome was closed. For many years there were mass protests against shipments of nuclear waste into the pilot conditioning plant and interim store which was also forced to close before it was full leaving wastes to be stored on sites at power stations around the country. Gorleben with its quiet rural hinterland of farming, waterland and forests has just the remnants of a nuclear industry. But, it was the very deep and relentless conflict over the site that brought about the eventual demise of not only Gorleben's nuclear industry but that of Germany's also. Thus, Gorleben stands at the far end of the continuum as the emblematic community where resistance proved sufficiently powerful to prevent nuclear development.

Each of the four communities reflects the first of the issues with which this book is concerned; the tendency of the legacy of nuclear power to be confined to its already established locations and the consequences this has both for the nuclear communities and for the management of nuclear wastes. I come now to my second theme, the tendency for these locations to express peripheral characteristics. They are places that can be described as 'peripheral' in several senses. In the next section I shall try to show how the concept of the periphery helps us to explore the nature of nuclear communities and to explain the processes and relationships which sustain their geographical location and persistence over time.

The nature of peripheral communities

The idea of peripheral communities was first put forward in a paper which I wrote with a colleague social scientist, Pieter Leroy, primarily to consider, as we put it; 'the sociological nature and political implications of the relationship between environmental quality and social inequality' (Blowers and Leroy, 1994, p.198). It was intended to apply to those places in which hazardous activities such as industrial chemicals, power stations, infrastructural works or waste facilities were located.

Such activities constitute what have come to be called 'locally unwanted land uses' (LULUs) (Popper, 1985) that must be located somewhere but which present environmental and health risks to local populations in existing or proposed locations. Nuclear facilities, including power stations but especially those activities concerned with decommissioning, clean-up and waste management, are, perhaps, the classic case of such LULUs. The central proposition posed in our paper was that LULUs tend typically to be located in already backward areas and that, therefore, their location reproduces and reinforces processes of 'peripheralisation'. I shall come to the latter idea in a moment. First, I need to outline the key characteristics of peripheral communities. In sum, they can be described as geographically remote, economically marginal, politically powerless and socially homogeneous. And they are also, whether in practice or in prospect, places of environmental hazard or risk. Those places considered in this book manifest all these qualities to greater or lesser degree as I shall endeavour to show.

Geographical remoteness

The idea of remoteness conveys being on the edge of the mainstream. It contrasts the periphery with the core or centre giving a sense of relative geographical isolation. That is not to say that a peripheral place or region is necessarily at the furthest distance from metropolitan centres, though that may be the case; rather, it is inaccessible in the sense of poor communications or distant in the sense of being beyond natural boundaries such as mountains or in different landscapes and landforms such as deserts, peninsulas or islands. Remoteness has connotations of being near or even beyond the frontier. In the case of nuclear facilities, remoteness has been a locational criterion dictated by the need for security in the case of military operations and safety as a precautionary principle in the early development of nuclear power. The idea that although nuclear power stations are inherently safe, precaution dictates they should be remote has survived, even in contemporary siting strategies. However, the definition of what constitutes remoteness and safety has remained vague and elusive. As Stan Openshaw (1986, pp.98–9) observed of the siting strategy for the early power stations in the UK:

> The problem was of course that no one had any real idea as to what an accident might involve and how remote a 'remote' site needed to be. It was all a matter of judgement and, since there were no nuclear disasters, it could well be claimed that the correct decisions were made; but it was certainly a gamble with only a vague idea of the public safety stakes.

Economic marginality

Peripheral communities are also marked by a general condition of economic dependence. They tend to be dominated by a single employer or sector privately or state-owned and dependent on that employer in terms of wages, investment and related service and sometimes supply chain activities. They may be described as one-industry or monocultural communities. The companies are often controlled from headquarters located elsewhere, sometimes even in foreign countries responding to regional, national and international ownership and interests. Likewise, investment decisions of state-owned enterprises must reflect wider national and public (often rendered as 'tax payer') interests. This does not mean that local interests are neglected, merely that they are not the only nor necessarily the determining interest taken into account by decision makers. In some cases where market and technological changes have resulted in restructuring, disinvestment and plant closure, peripheral areas may become abandoned and depopulated or dependent on state welfare. In this situation they are attractive to the investment and jobs that come with hazardous industries while also vulnerable to the attendant environmental risks. Nuclear communities, especially those with substantial employment like reprocessing facilities, tend originally to have developed in economically marginal areas, either where there was little pre-existing industry or competition or in places where they would be welcome to replace former activities that had disappeared or declined. Moreover, the nuclear industry has a high survival rate since, once established in specific places, it tends to continue its decommissioning, clean-up and waste management function long after the production has ceased.

Political powerlessness

The economic power exercised by dominant companies in peripheral communities may also be translated into political power. In the most obvious sense major companies will seek to influence governments at all levels to provide acceptable taxation and investment regimes and, where necessary, support and subsidies while, at the same time, ensuring the provision through public expenditure of adequate education for their workforce, health and social services and other services possibly including public transport, cultural and recreational facilities. Conversely, companies contribute to national and local taxation, undertake training of their workforce, pay wages and make other investments in the community. Company employees may participate directly in the governance of the community through election to councils or membership of community organisations. Through their contributions to community well being, companies are also serving their own interests as good employers and acceptable neighbours. On the other hand, companies may be perceived as ruthless and exploitative taking far more out of the community than they put in, acting as local agents for interests based elsewhere. Indeed, companies may not need to act at all since, merely by their presence, they can achieve what they need through their powerful reputation. A state of political stasis or even paralysis may obtain where a company refrains from acting since its interests are served best by non-decision making (Crenson, 1971) and a community likewise sees no point in acting in circumstances of 'mobilisation of bias' (Schattschneider, 1960). In short, power within a community is about the political context within which actions are taken and decisions are made; and the context is developed, shaped, manifested and sustained through the political relationships of negotiation, compromise and acceptance which constitute the culture of power in the community.

Although political power is formally exercised within communities by elected representatives such as mayors, councils, assemblies or other institutions, their powers are heavily constrained and circumscribed. Apart from the influence exerted by dominant companies directly or indirectly, locally elected bodies often have limited resources and powers and are subservient to regional and national political policies and decisions. In many countries the power of local authorities has been continually eroded as central government assumes more and more control over finance and functions. At the same time, in peripheral communities, local government may also be subdued by the demands and expectations of its dominant industry. It may be unwilling or unable to respond to NGOs which are able to mobilise people in opposing polluting activities and demanding a cleaner environment. Bereft of sufficient resources, besieged by opposing interests and undermined by external forces, a pervading condition of political powerlessness is likely to become a chronic condition in peripheral communities. Companies, though seemingly powerful within these communities, may themselves also feel beleaguered by public discontent and opposition within the community as well as neglect by distant owners or government. For both companies and councils there is the sense that key decisions are taken elsewhere. Unsurprisingly, a mutual sense of powerlessness leads inexorably to an embattled and united front against a hostile world and a reluctant acceptance of being on the periphery.

A culture of acceptance

The sense of economic and political subordination is also manifested in the social culture of these peripheral communities. The introduction, intrusion almost, of modern large-scale hazardous technology in remote regions disturbs and in some cases overwhelms and displaces pre-existing more traditional economic activities like farming and fishing. Over time it may be supposed that two tendencies develop that enable a transition to occur from one form of traditional community to another. One trend is the changing economic base marked by the growth of employment of a workforce of skilled and semi-skilled as well as professional workers. With this change comes a second tendency, the growing inequalities that develop between those benefiting from the wealth and incomes generated by the new industry and those remaining in traditional occupations, small companies, the service sector and those without work. Although such inequalities may be divisive, the separatist nature of a community on the periphery nurtures shared feelings of togetherness. There is often support for the company and what it means and brings to the area that extends beyond its immediate employees. This translates into such ambiguous feelings as fatalistic acceptance, resignation, defensiveness and a rather resentful pride in living in a community that is surrounded by an indifferent or even mildly hostile world. There is also a more positive, even aggressive belief in the role the community plays in bearing a burden of risk on behalf of the wider society. This solidarity finds political expression in defence of the industry against attack from environmental NGOs and explains why environmental groups are often absent or often unable to mobilise effective campaigns within peripheral communities. In the case of the nuclear industry this culture of acceptance, or 'nuclear culture' (Loeb, 1986) finds widespread expression and helps to explain the apparent willingness of nuclear communities to come to terms with the hazardous industry at their heart.

Environmental risk

A defining characteristic of peripheral communities is that they are places associated with environmental risk. They may be places of hazardous activities where chemical pollution, contamination or some form of physical degradation is taking place. Or they may be places of risk where there are no visible signs but where the possibility of serious hazards exists albeit with varying and sometimes vanishingly small probabilities of occurring. Nuclear facilities tend to come into this category, places where extremely dangerous materials are manufactured and stored and which routinely make discharges into rivers and the sea or emissions into the atmosphere. These emanations present no more than an 'acceptable risk' as determined by scientific experts, though the definitions, data and pathology upon which such assumptions are based are riddled with uncertainties and, therefore, controversial. There are uncertainties, too, about the safety and security of long-term storage of radioactive wastes and, some sites like Hanford and Sellafield discussed later in this book, are notoriously dangerous and problematic. Whatever the statistical calculation of risk from routine procedures, nuclear plants are also potentially vulnerable to low probability/high consequence risk of catastrophic incident whether accidentally or deliberately brought about. Major incidents have been fairly frequent (Sovacool, 2011; Sovacool and Valentine, 2012) and major accidents can and have occurred as those at Mayek (Russia, 1957), Windscale (UK, 1957), Three Mile Island (United States, 1979), Chernobyl (Ukraine, 1986) and Fukushima (Japan, 2011) testify. Indeed, the fact that such accidents, though infrequent, occur at regular intervals has led Charles Perrow to describe them as 'normal'. In the characteristic tight coupling of interdependent technological components and complex organisational systems of nuclear plant 'multiple and unexpected failures are inevitable' (Perrow, 1999, p.5). Communities living near nuclear plants are conscious of the possibilities of incidents and accidents but tend to internalise their anxiety and necessarily adopt an attitude of denial. The willing suppression of disbelief is greatly aided by the emphatic commitment to safety and security promulgated by the nuclear industry. But, while acceptance of environmental risk is a necessary condition of living in an established nuclear community, resistance by nascent nuclear communities to the prospect of such risk is the reason why the nuclear industry has found it so difficult to establish itself beyond its existing frontiers. In later chapters I shall consider the contrasts between a community intent on protecting itself from economic risk by supporting its nuclear activities such as Sellafield and one, Gorleben, that is determined to resist the environmental risks that a nuclear plant would bring.

Peripheralisation and power

Peripheral communities, then, are places of environmental risk in areas that are relatively isolated geographically and which tend to be characterised by economic dependence, political powerlessness and a culture of acceptance. There is, in these communities, a paradox of power. On the one hand they appear the powerless victims living in places of stigma, degradation and economic dependency. On the other, by their very presence as places which bear the risk and the legacy of nuclear energy, they possess a claim on the national resources for clean-up and the political leverage to ensure that their livelihood and continuing existence is supported. The nuclear peripheral communities have become places which have no choice but to host industries that are unwanted elsewhere. These nuclear communities on the periphery are created and sustained by processes of push and pull. There is the pull that they can exert for industries that provide the material basis for their existence. But, there is also the push that can be brought to bear by communities that have no wish to host risk-creating and polluting industries. Thus the power to attract is complemented by the power to repel in reinforcing and reproducing nuclear communities. In our paper we called this a 'process of peripheralisation' (Blowers and Leroy, 1994).

Peripheralisation is manifested in the nature of conflicts over the development and location of LULUs. Although most of the time these conflicts are latent and potential they may be brought to life when a proposal for a new nuclear facility is made. In some, relatively rare, cases there may be resistance to new facilities in already developing nuclear communities. As we shall see later in the book, the possibility of locating a geological disposal facility in West Cumbria, near to the Sellafield reprocessing works in the UK, met with both support and opposition in the area. More typically, resistance is strongest in areas with little or no experience of nuclear activities such as we shall see in the case of Gorleben in Germany. Even so, the power to resist varies greatly, depending on the social, economic and political circumstances and the geographical context. In those areas where population may be thin or where there has been economic decline and deprivation there may be little resistance and, in a few cases, the nuclear industry has been able to establish itself in greenfield locations. An example, covered in Chapter 4, is Bure, in France, the location for the prospective national deep geological repository.

More common are those areas which are able successfully to mobilise against the introduction of new nuclear facilities. They may be peripheral in location but are rarely peripheral in other senses, having diversified economies, communities that are socially heterogeneous with leaders and elites able to mobilise resources of protest and apply political pressure. In such communities, campaigns of protest led by local NGOs and political leaders and backed by national organisations are capable, at moments of threat, of developing and delivering united opposition that cuts across conventional divides of class, party and culture. At such moments, it can prove possible to reveal and defend those more traditional patterns of social integration comprising the sense of community which, though usually dormant, can be aroused when an external threat to its integrity is posed. While the numbers actively engaged or even interested need not be large, evidence of widespread, if passive, support gathered through media campaigns, opinion surveys, petitions or other methods is necessary to command attention and achieve success. The power of resistance is, in some cases, strengthened by the ability to mobilise coalitions not just within communities but between them when several locations are threatened simultaneously. There are also instances where the identification of sites for hazardous activities is initiated generically but where communities, fearful of being chosen, take action to forestall the possibility of eventually being selected. Instances of these various possibilities are encountered in later chapters in this book. The point to make here is that the process of peripheralisation involves the mobilisation of resources of power that both repel and attract industries that are associated with risks to environment and health. It is the inequality in resources possessed and mobilised by those communities able to resist, combined with the powerlessness of communities that have little choice but to accept, that ultimately and inevitably confirms a pattern of location of hazardous industries in peripheral locations.

We expressed this outcome, rather portentously, as our central thesis in our paper, that 'LULUs tend typically to be located in already backward areas and that, therefore, their location reproduces and reinforces processes of "peripheralisation"' (Blowers and Leroy, 1994, p.198). That is not to say that peripheral communities can be simply conceived as benighted places, left stranded by modern society, places of danger, economically precarious, socially deprived. Certainly they may possess some of these features but the condition of peripherality is more subtle and complex. It is important not to take generalisation too far since, as we shall see, each place expresses the conditions of the periphery in specific ways. What, perhaps, distinguishes them is the interdependent relationship between industry and community which is, at once, both unifying and divisive. The established nuclear communities I shall explore possess a sense of integrity, of common interest and purpose in sustaining their economic base. This is manifested in the strength of solidarity and integration that resists external threat. But, there is also the internal divisiveness caused by the inequalities in income and wealth that a dominant employer promotes. Together company and community are able to exert leverage by virtue of the burden they bear and the role they undertake as guardians of the nuclear legacy on behalf of the wider society. Though not powerful, neither are they powerless. And, whatever the fortunes of the nuclear industry, those places that guard its legacy and which are recorded here will continue to be supported for decades to come.

Discourses of nuclear power

The thesis of peripheralisation helps to explain how nuclear communities come to be established where they are and the social and economic conditions that account for their continuing persistence. It brings me to my third theme, focusing on power relations, in the sense of the inequality in power resources between places that creates and perpetuates a pattern of uneven development whereby hazardous activities end up in peripheral locations. The geography of the nuclear industry, once established, has been relatively stable, yet this locational stability masks a history of nuclear energy that has been anything but stable. Nuclear communities have experienced the economic vicissitudes and uncertainty that accompanies dependence on a controversial industry whose fortunes have waxed and waned according to changing political priorities for energy security and environmental protection. These changes impact on nuclear communities and can be explored in terms of power relations. The interaction, confrontation and diffusion of values, beliefs, principles and prejudices involved in power relations constitute what are often referred to as 'discourses'. Discourses are ways of apprehending the world, shaping, defining and constraining the way it is understood and informing the actions and choices we take. Maartin Hajer (1995, p.44) has provided a formal definition for discourses as 'a specific ensemble of ideas, concepts and categorisations that are produced, reproduced and transformed in a particular set of practices and through which a meaning is given to physical and social realities'.

This idea of power as discourse produces the context within which mobilisation of power as resources occurs. Discourses enable and constrain the ability of groups within society or communities or industries to assemble and deploy resources of power such as investment power, technological power, power to protest and so on. In other words they set the frame of reference of possibilities within which power relations interact. Put in terms of the nuclear industry, changing discourses have shifted the terms of engagement and the availability and strength of resources in a continuing conflict between those interests favouring the development of nuclear energy and those opposed to it.

It is possible to distinguish four successive discourses that have shaped the power relations that have determined the development of the nuclear industry. Although distinguishable, they are not distinctively different. Each contains similar elements to the others but with different degrees of emphasis or tendency. Ian Welsh (2000) tends to emphasise the continuities he found inherent in 'the points of origin, repetitions and reformulations which constitute nuclear discourse over a period of fifty years' (p.15). Nevertheless, it seems clear that discourses shift over time as one passes to another but, for a while, they may coexist and overlap until one succeeds another and, for a time, becomes dominant. Thus there are periods when discourses appear to be settled and times of transformation when new concerns and issues unsettle and eventually displace the dominant discourse. While the details and, to some extent, the timing of these discursive shifts has varied in different countries, the pattern has, by and large been repeated everywhere in Western democratic contexts. Changing discourses have influenced power relations and the deployment and withdrawal of nuclear facilities with profound consequences for the nuclear communities and the legacy they manage. The first three discourses are reasonably easy to distinguish in terms of their influence on the power relations affecting the four communities that are the focus of this book. First, there is what I have termed the 'Discourse of Trust in Technology' which coincided with the emergence and early development of nuclear power. The following 'Discourse of Danger and Distrust' covers the period of conflict between the rapidly expanding nuclear industry and increasing concerns about its environmental consequences. Horace Herring (2005) warns against making too sharp a distinction between a 'golden age of public acceptance' and a subsequent anti-nuclear phase pointing to the evidence of opposition to nuclear power that occurred even during its early period (p.68). It is possible to identify a third phase, the 'Discourse of Consensus and Cooperation', marked by a decline in nuclear development in many countries which opens up opportunities for conflicting forces to cooperate in the management of the nuclear legacy. Finally, it is possible to discern a contemporary 'Discourse of Security', immanent in some countries experiencing a revival in nuclear's fortunes but taking a different turn in others, like Germany, where nuclear energy is in continuing decline.

Discourse of Trust in Technology

The original purpose of the nuclear industry was to produce weapons capable of mass destruction of environment and life on an unprecedented scale. Its terrifying possibilities were unleashed in the atomic bombs that destroyed Hiroshima and Nagasaki within three days in August 1945. In the post-war world the military purpose of the nuclear industry expanded as the United States, soon joined by the Soviet Union, the UK and France, built reprocessing works and bomb assembly plants producing atomic and hydrogen warheads intent on the mutually assured destruction that would occur if ever they came to be used. In parallel there developed programmes using nuclear fission for civil purposes, the antidote to its warlike origins. The peaceful atom appeared as a technological wonder to a world struggling from war in conditions of austerity (Kynaston, 2007). Nuclear power would be cheap (too cheap to meter), clean (safe) and would provide unlimited electricity. Its downsides, potential accidents, problems of waste management, escalating costs, technological failures, were ignored or set aside in the enthusiastic embrace of a nuclear utopia.

The nuclear industry became the perfect exemplar of the discourse of Trust in Technology. It was characterised by a confidence in scientific discovery and its application in modern technology in a world that trusted and deferred to expertise. It was an unquestioning faith and, in truth, the secrecy which shrouded nuclear activities made questions difficult to pose, let alone answer. This was the period in which the nuclear reprocessing works such as Hanford and Sellafield which are discussed later in this book were developed, initially for military production but later for power plants, nuclear fuel manufacture and experimental reactor technology. Programmes for nuclear power for electricity production began to develop and, by the 1960s, power stations began to appear in the UK, United States and elsewhere though the rapid expansion in most countries came a little later, during the 1970s. Overall, for the first three decades after the war, the nuclear industry was carried forward on a wave of enthusiasm and optimism. Although opposition was not entirely absent it was fleeting and typically directed at specific projects, usually nuclear power stations; a broader anti-nuclear movement had yet to materialise. The production of nuclear energy had managed to emancipate itself from its darker military connections. Although major accidents occurred (both Mayak in the Urals and Windscale in the UK occurred in 1957), they were covered up and made no impression on the upward curve of nuclear development. During this period of Trust in Technology nuclear communities emerged in their remote and peripheral locations. This was a period when the nuclear industry was fostered by a powerful combination of government support and scientific expertise. This was, in Welsh's words, a period of 'peak modernity', 'a "moment" during which the will to back heroic scientific projects intended to modernise the world existed among the leaders of both democratic and socialist states' (2000, p.18).

Discourse of Danger and Distrust

Gradually, from the late 1960s onwards, the power relations began to change as the nuclear industry became more vulnerable to a growing opposition. There are three main causes for the declining power of the nuclear industry during the last quarter of the twentieth century. First, there was the decline in the rate of expansion of the nuclear industry itself, uneven and slow at first, then more rapid as it passed its peak in construction and development in Western countries. As a measure of its recession, in 1975 there were 156 reactors on order worldwide; by 1979 there were only 102 and by 1980 the number had fallen to 60. Of course, as more and more power stations were commissioned so the output of electricity from nuclear energy continued to increase thereby adding to excess electricity generating capacity which was a major factor in the decline of orders. In addition, the need for energy security that had spurred the expansion after the oil crisis in the 1970s, notably in France, had lost its impetus as nuclear energy was increasingly unable to compete with fossil fuels in the 1980s. High capital costs, cost overruns and construction delays all caused nuclear to lose favour.

A second cause of nuclear's declining political power was the growing awareness and concern about the problems of the back end of the nuclear cycle, reprocessing and nuclear waste. During nuclear's halcyon days the rapid deployment of the new technology subsumed all other issues. Indeed, there was little to trouble the industry, at first, since there was not much waste and what there was had been unceremoniously dumped in the sea or tipped into tanks and silos at major nuclear sites. The problems of decommissioning and clean-up were far in the future. Nevertheless, the problem of what to do with nuclear's growing legacy of waste had become a matter of public and political interest by the 1970s thus diverting some of the glamour from the nuclear enterprise. In those countries with a nuclear reprocessing cycle there was increasing concern about its costs and justification in the face of a diminishing market for plutonium. The early promise of deployment of fast breeder reactors was fading as problems with developing the technology were encountered. As the pace of nuclear development slackened so, inevitably, the potential demand for mixed oxide fuel using recycled plutonium in reactors also fell away. With the ending of the Cold War and arms reduction, military reprocessing for plutonium diminished. In the four countries considered in this book, reprocessing for civil purposes had varying fortunes. It was abandoned in the United States at the end of the 1970s, failed to get started in Germany and continued in the UK increasingly as a means of waste management; only in France with its strong commitment to nuclear energy was reprocessing an integral part of the nuclear fuel cycle. Elsewhere,[2] as once-through nuclear cycles were almost universally adopted, spent fuel was stored rather than reprocessed. Together with the high level wastes from reprocessing, spent fuel constituted the most dangerous and difficult materials of the legacy of nuclear power.

Thirdly, the dangers posed by nuclear energy were spectacularly demonstrated by its propensity for catastrophic accidents, the third reason for nuclear's decline during this period. The accident at Three Mile Island in 1979 severely undermined trust in the nuclear industry and precipitated a shift in power relations that had already been occurring. The cataclysmic melt down at Chernobyl seven years later provided the ultimate confirmation that nuclear energy was, indeed, a dangerous technology with potentially global implications. Chernobyl fell 'like a hammer-blow on an industry that was already groggy' (Hawkes et al., 1986, p.222).

The discourse of Danger and Distrust provided fertile space for the emergence of anti-nuclear opposition. The end of the 1960s was an era of palpable social change from the conformist, deferential society in awe of technological progress of the post-war years. New ideas were emerging and new social movements developing which began first to erode, then undermine and finally to supplant and replace the trust, optimism and certainties which had carried the nuclear enterprise forward hitherto. In particular there were trends within civil society which challenged the established order and sought to open up closed processes to more democratic participation. Environmental groups typically took a lead for, as Goodin observes, 'the environmental area has led all other issue areas in democratic innovations' (2003, p.164). The degradation of environments caused by pollution and resource exploitation provided visible and easy targets in the form of dirty industries, big companies and inadequate governance. They presented targets around which it was possible to mobilise powerful alliances cutting across political, class and geographical boundaries.

The peace movement had arisen to oppose the development and deployment of nuclear weapons. Somewhat separate but obviously related anti-nuclear protests began initially targeting nuclear power stations. Later, as the industry developed its back end functions so the scope of opposition and protest widened to incorporate nuclear reprocessing and especially nuclear waste. In many countries there arose anti-nuclear movements coordinated at national level, linked internationally and with grass roots support often based in local organisations at nuclear sites. Environmental organisations proved particularly adept at mobilising protests and political pressure in cross-cutting combinations. They focused attention on the nuclear industry more generally but also inspired actions and provided support to local campaigns against specific projects. As their power and influence rose so they helped to promote a discursive shift. In particular they became successful in preventing, or at least impeding, the nuclear industry from breaking out into new locations as we shall see in the case of Germany and the UK. Thus, a combination of a weakening industry and a strengthening opposition transformed the nuclear discourse and power relations thereby ensuring that the nuclear industry, by and large, was restrained and confined to its etablished peripheral locations.

Discourse of Consensus and Cooperation

Towards the end of the last century, a new nuclear discourse was emerging. By contrast with the discourse of Danger and Distrust with its tendency to polarised conflict and political turbulence, a more tranquil period in power relations between nuclear and anti-nuclear perspectives ensued. The pace, timing, depth and characteristics of the process of discursive transition varied from country to country responding to differences in systems of governance, political culture and the circumstances of the nuclear industry. But, to a greater or lesser degree, and for a relatively short period, there ensued what I will term a discourse of Consensus and Cooperation marked by a coming together on all sides in an effort to find solutions for the nuclear industry's intractable legacy. This discourse may only have been an intermission and, in the UK for instance, it did not supplant the fundamental basis of power relations which were able to resume later. By contrast in Germany the shift in power relations continued leading ultimately to the demise of the nuclear industry for a variety of reasons.

The shift was, in part, a response to the decline of the nuclear industry. By the turn of the century it seemed clear, certainly in the four countries covered in this book, that any further expansion in nuclear energy was over, at least for a time. Although the basic reason for the slow down was its economic cost and uncompetitiveness, nuclear energy had lost political favour as a consequence of its market failure, public concern and the increasingly confident pressure from its opponents. The new discourse opened up space providing politial opportunity for greater interaction and potential cooperation between hitherto contending interests. Opponents of nuclear energy could now direct their attention to the problem of nuclear's legacy, in particular what to do with its accumulating waste. The nuclear industry was perforce encouraged to clean up its act. Moreover, the conditions of engagement had changed. From being antagonists, both sides in the nuclear conflict now found they were, at least in principle, on the same side as protagonists when it came to ensuring the safety and security of the management of radioactive wastes. It would be wrong to exaggerate this coming together since though there was broad agreement on the ends (safe management), the means of its achievement (storage or disposal) were far from clear and remained contestable. And, there was inherent conflict between the pro-nuclear interests for whom the end was the means to justify future expansion and the anti-nuclear movement who perceived the end as the burial of the nuclear industry itself.

The broader societal and political circumstances for a cooperative relationship between erstwhile opposing factions were propitious. The new discourse reflected the so-called 'participative' or 'deliberative' turn which stood in contrast to the more elitist, top down, reactive mode of policy making by which governments traditionally retained the initiative in policy making and control over decisions. The new discourse became associated with a more democratic and inclusive approach to policy and decision making especially in the environmental field. Cooperative forms of environmental governance began to develop characterised by the search for agreement or consensus on policy processes and outcomes (Glasbergen, 1998; Leroy and Verhagen, 2003). Environmental politics and policy making explored means of integrating participative and representative forms of democratic decision making. Through an increasingly sophisticated array of methods of public and stakeholder engagement (PSE), a veritable industry of participative democracy developed. The new approach, based on commitment to 'engagement', 'openness and transparency', dialogue and deliberation was, potentially, quite transformative in giving environmental groups, citizens' based organisations and the wider public a greater say in decision making. These new democratic forms performed two vital functions in decision making. One, by encouraging wider participation, they helped to ensure greater public acceptability and confidence in decision making. Two, by providing evidence of citizens' and stakeholders' involvement and support, these new approaches facilitated the legitimation of policy through representative decision making.

Deliberative approaches to decision making found fertile territory for expression in the nuclear industry and especially in policy making for radioactive waste. Although waste had long been recognised as the Achilles heel of the nuclear industry, efforts to find publicly acceptable, scientifically credible and long-term solutions had foundered in every country. The nuclear industry, supported by governments, had adopted deterministic strategies, by which methods of management and sites for facilities were decided and then announced to an unsuspecting public and defended against inevitable opposition before being abandoned (an approach that acquired a familiar acronym, DAD which, after successive retreats could be extended as DADA, Decide Announce Defend Abandon). These failures had largely inhibited the nuclear industry from breaking out of its bastions into greenfield territory in its search for suitable locations for nuclear waste repositories. There were exceptions; in the United States the WIPP (Waste Isolation Pilot Plant) facility in a greenfield location in New Mexico was opened in 1999 and in France, Bure became the site for an underground laboratory, the likely site for a national repository (see Chapter 4). But in the US, Yucca

Mountain, Nevada, was a victim of a stalemate between contending political forces and the site at Gorleben in Germany which had been the focus of mass protests for three decades was withdrawn, though not completely abandoned (see Chapter 5). In the UK, the new discourse opened up discussion and developed a cooperative and voluntarist process of site selection which stalled for lack of political consensus. Only in Finland and Sweden were acceptable and suitable locations for a deep repository found and they were at existing nuclear locations. While the discourse of Cooperation and Consensus had opened up the issue of radioactive waste management, progress towards solutions remained mired in controversy, procrastination and impasse. In any event, in the early years of the new century there were evident signs that the cooperative discourse was on the wane and the possibility of a return to a more *dirigiste* approach had set in.

Discourse of Security

Viewed over the longer term the discourse of Cooperation and Consensus appears both short-lived and exceptional. In fact, it probably did not get sufficiently established everywhere to constitute anything approaching a dominant discourse. Certainly, it achieved, in some countries and for some areas of environmental policy making, at least a transitional, if not transformative stage sufficient to ensure there could be no return to the former approaches to decision making. Once the cornucopia of participative democratic methods had been revealed it could not be denied. Thus, at the very least, the discourse was interwoven with its predecessor creating a hybrid which, flourishing briefly and in specific circumstances and policy areas, opened up new possibilities and approaches that cannot be gainsaid. But, the discourse possessed a certain optimism about the possibilities of consensus that flourished for a while in relatively quiescent circumstances. Very soon the mood had shifted towards a more pessimistic and apprehensive view of the world and new concerns took hold, focused around issues of security.

Although security is always a preoccupation for governments, a discourse of security began to emerge in the early years of this century which transcended various theatres of strategic policy making. Four components of this globalised discourse may be discerned. Perhaps the most obvious manifestation was a concern with what may be called 'securitisation' (Aradau, 2009), the emphasis on physical protection of the state and its citizens from terrorist attacks following 9/11 in 2001 and subsequent conflicts especially in the Middle East leading to a heightened sense of threat and increasing surveillance and controls. A second component, developing a little later, was a focus on economic security in the wake of the global recession following the collapse of the banking system in 2008. Partly provoked by economic and political instability was the third component, an increasing concern for energy security. Fears of energy shortages and dependence on unstable regimes for oil and gas supplies encouraged moves towards greater national self-sufficiency. The overdependence on consumption of fossil fuels became the focus of the fourth component, the emphasis on environmental security in an era of climate change.

The nuclear sector fits well with the security agenda. The endemic tendency towards secrecy in nuclear decision making has been reinforced by the need to protect nuclear plants from terrorist attacks. In terms of economic security nuclear is portrayed by some governments as a cost-effective indigenous and low cost option, though its economic performance relies on subsidies and price guarantees. In terms of energy, nuclear is said to provide security as part of the energy mix offering base load electricity and low dependence on foreign imports. When it comes to environmental security, nuclear energy has been seized on as a supposedly low carbon form of electricity production contributing to mitigating the problem of climate change. On the twin planks of energy and environmental security nuclear energy began to stage a revival in the early years of the century. The so-called 'nuclear renaissance' has stuttered in the West and, in some countries, notably Germany, nuclear phase-out has been reinforced in the aftermath of the disaster at Fukushima Dai-ichi (Hindmarsh, 2013). Progress worldwide has since been slow. Of the 66 nuclear power stations under construction in 2015, over half were in China (23), Russia (9) and India (6) with five in the United States and only two in Western Europe (at Okiluoto in Finland and Flamanville in France), both suffering cost and time overruns and projects such as Hinkley Point in the UK encountering financial and regulatory problems. Nevertheless, nuclear's revival helped instigate the shift in power relations that came with the new discourse of security. The nuclear industry readily grasped the political opportunity created by anxiety over security issues. Anti-nuclear groups, emphasising the threats to security posed by nuclear energy, in some countries revived their combative mission to prevent the development of new power stations and other nuclear facilities including waste management projects. Their attack was based on the insecurities that nuclear power implied; insecurity from a terrorist attack or from a major incident; economic insecurity from high costs and the need for subsidies far into the future; and environmental insecurity brought about by the pollution, contamination and potental devastation associated with nuclear energy and its legacy of nuclear waste. Their impact varied according to the state of the political discourse; it was strong in Germany, influential in France, and weaker though not ineffectual in the US and UK.

It may be argued that, in the case of nuclear energy, the discourses of Security and Cooperation and Consensus are coexisting to form a hybrid discourse. In so far as it has developed, the Security discourse has encouraged the revival of approaches to policy making redolent of earlier periods and, consequently, weakened the more participative and democratic approach of the discourse of Cooperation and Consensus. In particular, there is a perceptible reversion to centralised decision making with a less participative and less inclusive style of governance though it would be wrong to exaggerate the change since participatory processes have become an integral element in decision making especially for major infrastructures such as geological repositories. The impact of this discursive change on policy making will vary according to the institutional and political arrangements in different countries as we shall see in the four cases examined in this book.

Nuclear oases – peripheral but permanent

Approaches to this study

So far, I have tried to establish that the legacy of nuclear power is a social as well as material concept. In material terms the legacy consists in growing accumulations of radioactive wastes, unwanted and potentially dangerous, in the form of derelict buildings and contaminated areas, solid and liquid waste materials held in storage tanks and dumps, abandoned experimental projects, shut down reactor cores, in sum the whole paraphernalia that signifies the end of operations. And yet, unlike many industries, nuclear's legacy is long lasting, almost eternal, requiring maintenance, management and vigilance into the far future. The processes of decommissioning, clean-up and radioactive waste management require a continuing commitment and place a burden of cost, energy and risk on the communities that must bear responsibility for looking after the legacy. It is these communities which are the central interest of this book; it is their experience and the implications it has for how we deal with nuclear's legacy now and in the future that has troubled and intrigued me throughout the years. I think I was one of the first to pronounce that nuclear waste is a social as well as a scientific problem. So, I have long felt it an obligation to try to set out to understand what this means and to fathom the consequences of this understanding.

I have approached this quest from the local perspective. I have done so for several reasons. Most studies tend to take a synoptic view, a top down perspective grounded in national concerns and generic approaches for application to local and specific circumstances. I have decided to foreground the local, to take a more bottom up, grass roots perspective considering the impact of broader processes and imposed policies on the lives and prospects of nuclear communities. This approach also fits with my own experience as a county councillor in Bedfordshire for nearly three decades during which time I chaired the Environment Committee. It was in the autumn of 1983 (a memorable time for nuclear politics in the UK as I shall show later in Chapter 3) that Nirex (the Nuclear Industry Radioactive Waste Executive) proposed a nuclear waste facility at Elstow just south of Bedford where I live and I was one of the leaders in the ultimately successful campaign to stop the project (Blowers et al., 1991). But, the experience had excited my interest in the problem of nuclear's legacy and I continued my active participation in policy making as, variously, a member of the UK government's Radioactive Waste Management Advisory Committee (RWMAC) from 1991–2003, the Committee on Radioactive Waste Management (CoRWM) from 2003–2007 and as a non-Executive Director of UK Nirex (2000–2004), the very company I had opposed as a county councillor in Bedfordshire. In each of these roles I brought a social science perspective to bear on the understanding and formulation of policy. Once I had retired from politics, and no longer a government adviser, I set up a local NGO, Blackwater Against New Nuclear Group (BANNG) to oppose proposals for new nuclear power reactors at the coastal Bradwell site in my home county of Essex, just up the east coast from London. Finally, I have acted as Co-Chair of the UK government/NGO Nuclear Forum which brings together NGOs, many of them based in local communities.

This local focus has inspired both my academic and political careers. As an academic I have tried to analyse theoretical concerns using empirical knowledge and experience in local political contexts. My political experience fed into an academic interest in democratic forms of decision making, especially the relationship between local and national levels. Again, I was able to explore my academic interest in patterns of inequality which led to an eponymous Open University course (Open University, 1976; Blowers and Thompson, 1976) in combination with a political concern for dealing with its social manifestations in my own political constituency in inner Bedford. As a leading member of a County Council I was able from direct experience to gain insights into the operation of power relationships in practice. Over the years these three concepts – democracy, inequality and power – became the focus of my intellectual imagination and my political commitment. My original disciplines of Geography and History provided me with a lifelong fascination with how and why places and communities develop where and in the way they do, and a recognition that understanding can only be achieved through knowledge of the places themselves. Thus, as should be evident from the earlier books I have written (e.g. Blowers, 1980, 1984; Blowers et al., 1991), I have used the material to hand, in the places I have represented in politics or come to know often as a participant in policy making or through research visits at home and abroad. My academic interests combined with my own empirical experience have, I believe, helped me to inculcate a social science interdisciplinary approach to an understanding of environmental politics, policy and power expressed in local contexts. This book is, to that extent, in the neglected tradition of community studies, a tradition which had its golden age long ago but which I hope in a small way to revive (Bell and Newby, 1971).

This book has been long in the making, in some ways too long for its conception and development have spanned changes in academic preoccupations and political priorities. This may lend the work a somewhat eclectic, even idiosyncratic focus, at times a kind of potpourri scattered over a wide terrain of thought and action. It draws on a wide range of sources and academic debates but without specific commitment to any one perspective nor making any pretence to engaging systematically with established theoretical protocols. Thus, it touches on various literatures but often tangentially and incomprehensively – to do otherwise would be a daunting and, for me, impossible task. In case this sounds like an *apologia pro vita sua* let me state where I think the strength of my approach lies. First, is my attempt at integration of a range of dimensions; of the theoretical and empirical, the academic and the political, the scientific and the social, the global and the local. Second, I have attempted to achieve this through a bottom up, local approach as a way of gaining insights into issues of inequality, power and politics that have wider relevance. Third, I have been able draw on a wide variety of source material that has helped to enliven and illuminate the local contexts I have studied including academic monographs, national and local government policy documents and reports, publications from local media, NGOs and the nuclear industry and, in addition, I have had access to a variety of unpublished, informal materials gathered online and through personal communication. Fourth, my lifelong involvement in local politics, government advisory bodies and environmental groups has helped me to apply a focus on the possibilities of policy making that is democratic and equitable from the perspective of a participant observer. Fifth, I have studied four specific nuclear communities as the case studies for a historical and geographical setting through which to gain insights into social processes working in a specific place over time. Finally, it is through visiting these places several times and talking with the people living there, especially community leaders, those working in the nuclear industry and those representing community interests, that I have tried to build up my picture of local communities and nuclear waste.

Nuclear communities, as I have defined them for this study, are the places of nuclear's legacy. They are the places where the end stages of the nuclear industry, its reprocessing, its decommissioning, its accumulated wastes and its clean-up are located. They are places on the periphery, differentiated by shared characteristics of remoteness, economic marginality, powerlessness and cultural distinctiveness. Each of the places covered in the following chapters has a different story to tell. In Hanford and, to a large extent in Sellafield, clean-up has become the *raison d'être* and long-term objective sustaining stable relations between the industry and community alike. In La Hague reprocessing still supports a culture of comfortable coexistence between industry and community while at Bure the social context has yet to develop. In Gorleben a long and embattled history has bequeathed a culture of conflict in which nuclear's legacy has been curtailed. Yet, while each story is, indeed, different it is possible to draw some broader conclusions about the relationships and prospects of communities and their implications for dealing with nuclear's legacy in the long run. That I shall attempt to do in the final chapter. For the present let the stories begin.

Notes

1 Possible exceptions to this are the Waste Isolation Pilot Plant in New Mexico, USA, developed in a salt formation for the reception of transuranic wastes which has been subject to suspension; and the deep repository called Onkalo in Finland currently under construction.

2 Reprocessing remains an integral part of the nuclear fuel cycle in Russia and in India. Japan is also committed to reprocessing but the opening of its plant at Rokkasho has been long delayed.

References

Aradau, C. (2009) 'Climate emergency: is securitisation the way forward?' In Humphreys, D. and Blowers, A. (Eds), *A Warming World*, DU311 Earth in Crisis, Milton Keynes, The Open University.

Bell, C. and Newby, H. (1971) *Community Studies*, London, George Allen and Unwin Ltd.

Blowers, A. (1980) *The Limits of Power: The Politics of Local Panning Policy,* Oxford, Pergamon Press.

Blowers, A. (1984) *Something in the Air: Corporate Power and the Environment,* London, Harper and Row.

Blowers, A. (1999) 'Nuclear waste and landscapes of risk', *Landscape Research,* 24,3, 241–64.

Blowers, A. and Leroy, P. (1994) 'Power, politics and environmental inequality: a theoretical and empirical analysis of the process of "peripheralisation"', *Environmental Politics,* 3, 2, Summer, 197–228.

Blowers, A. and Thompson, G. (Eds.) (1976) *Inequalities, Conflict and Change,* Milton Keynes, The Open University Press.

Blowers, A., Lowry, D. and Solomon, B. (1991) *The International Politics of Nuclear Waste,* London, Macmillan.

Bradley, D.J. (1998) *Behind the Nuclear Curtain: Radioactive Waste Management in the Former Soviet Union,* Columbus, Ohio, Battelle Press.

Brown, K. (2013) *Plutopia,* Oxford, Oxford University Press.

Crenson, M. (1971) *The Un-Politics of Air Pollution: A Study of Non-Decisionmaking in the Cities,* Baltimore, The Johns Hopkins Press.

Gerber, M. (1997) *On the Home Front,* Third Edition, Lincoln and London, University of Nebraska Press.

Glasbergen, P. (Ed.) (1998) *Co-operative Environmental Governance,* Dordrecht, Kluwer Academic Publishers.

Goodin, R. (2003) *Reflective Democracy,* Oxford, Oxford University Press.

Hajer, M. (1995) *The Politics of Environmental Discourses,* Oxford, Oxford University Press.

Hawkes, N., Lean, G., Leigh, D., McKie, R., Pringle, P. and Wilson, W. (1986) *The Worst Accident in the World,* The Observer, London, Pan Books and William Heinemann.

Herring, H. (2005) *From Energy Dreams to Nuclear Nightmares: Lessons from the Anti-Nuclear Power Movement in the 1970s,* Charlbury, Oxon, Jon Carpenter.

Hevly, B. and Findlay, J. (1998) *The Atomic West,* Seattle and London, University of Washington Press.

Hindmarsh, R. (Ed.) (2013) *Nuclear Disaster at Fukushima Daiichi,* New York and London, Routledge.

Kynaston, D. (2007) *Austerity Britain 1945–51,* London, Bloomsbury Publishing.

Leroy, P. and Verhagen, K. (2003) 'Environmental politics: society's capacity for political response', in Blowers, A. and Hinchliffe, S. (Eds) *Environmental Responses,* Milton Keynes, The Open University and Chichester, John Wiley.

Loeb, P. (1986) *Nuclear Culture,* Philadelphia, New Society Publishers.

Open University (1976) *Patterns of Inequality,* Open University Course D302, Milton Keynes, The Open University.

Openshaw, S. (1986) *Nuclear Power: Siting and Safety,* London, Routledge and Kegan Paul.

Perrow, C. (1999) *Normal Accidents: Living with High Risk Technologies,* Chichester, West Sussex, Princeton University Press.

Popper, F. (1985) 'The environmentalist and the LULU', reprinted in Lake, R. (Ed.) (1987) *Resolving Locational Conflict,* pp.1–13, New Brunswick, NJ, Rutgers University Press.

Schattschneider, E. (1960), *The Semi-Sovereign People: A Realist's View of Democracy in America.* New York, Holt, Rinehart and Winston.

Schlosser, E. (2013) *Command and Control,* London, Allen Lane.

Sovacool, B. (2011) *Contesting the Future of Nuclear Power,* Singapore, World Scientific.

Sovacool, B. and Valentine, S. (2012) *The National Politics of Nuclear Power,* London and New York, Routledge.

Welsh, I. (2000) *Mobilising Modernity: The Nuclear Moment,* London and New York, Routledge.

2

HANFORD, USA

An enduring legacy

This project symbolises our strength as a nation ... our desire to end wars and preserve peace, to build rather than despoil. To have permitted this resource to be wasted would have been in conflict with all the principles of resource conservation and utilisation to which we are committed.

(President John F. Kennedy at the groundbreaking ceremony for the N reactor at the Hanford site, Washington State, 26 September 1963)

If future generations are to remember us more with gratitude than sorrow, we must achieve more than the miracles of technology, we must also leave them a glimpse of the world as it was created, not just as it looked when we got through with it.

(Lyndon Baines Johnson, quotation on information board, Bryce Canyon, Utah)

Introduction

In eastern Washington state lies the mid-Columbia valley, a semi-desert region of bare and barren brown and yellow hills and plains of sagebrush interspersed with verdant, irrigated valleys of orchards, vineyards, cattle pastures and rolling wheatlands. In the midst of this area, where the Columbia is joined first by the Yakima and a few miles further downstream by the Snake River, lie the cities of Richland, Kennewick and Pasco which together make up the Tri-Cities with a population of more than 200,000. But back in 1942 there were only tiny settlements here based mainly on ranching and irrigated croplands in the wind blown sagebrush desert. The Lewis and Clark expedition had passed through here in 1805 (DeVoto, 1953) and pioneers had gradually settled in areas occupied by Indian tribes who lived off fishing and gathering. This was the American West, a region where environmental hazards and economic risk made life unpredictable for the early pioneers (Raban, 1997). The railway which reached Pasco in 1884 had opened up the area which during the subsequent half century experienced the booms and

FIGURE 2.1 General Leslie R. Groves, director of the Manhattan Project, at Hanford on 29 September 1945

Source: US Department of Energy

FIGURE 2.2 Hanford under development during the Second World War

Source: US Department of Energy

depressions inherent in its dependence on agricultural markets elsewhere. Despite the precariousness of life in the Columbia Basin those who made it their home 'felt a bond to their windy expanses, and their roots grew as deep and tenacious as those of the sagebrush' (Gerber, 1992, p.22). By the time of the Second World War this sense of identity with the rhythms of rural life was about to be uprooted and the economy of the mid-Columbia region was soon to be utterly transformed.

The history and causes of this transformation are the subject matter of this chapter. The research is based on four visits to Hanford (in 1987, 1999, 2004 and 2013) and interviews conducted with community representatives during the latter three visits. A list of those interviewed is given at the end of the chapter. Here, as in the three chapters that follow, I shall explore the three themes outlined in chapter 1. In the next section, I shall briefly set the scene of Hanford's nuclear legacy, a legacy that will continue to have a profound impact on the environment and community into the far future. This establishes the context for the following section which takes a historical perspective examining the shift from an economy based on nuclear production to one where the mission is environmental clean-up. This opens up the issue of how far and in what ways the political and social context of Hanford both influenced and responded to the changing nuclear discourse. In the third section of the chapter I shall explore the idea and meaning of a 'peripheral community' in a specific context. Hanford offers some interesting insights into the dynamic nature of the concept of peripherality. It exemplifies how, despite the loss of its original mission, Hanford has been able to sustain a regional importance and a wider national role. Hanford provides an enticing case study in the power relations which shape the political geography of nuclear communities.

Hanford – the legacy

As I write this more than 70 years later, the turbulence of the war years when Hanford was at the heart of the Manhattan Project has long passed. For Hanford was the place where the plutonium for the bomb that exploded over Nagasaki was made. The long aftermath of the Cold War when Hanford maintained its role in plutonium production for the nation's arms race and the subsequent period of massive clean-up have also left their mark, both visible and invisible, on this compelling landscape. I have visited all parts of the site though much of it has restricted access. But it is possible to grasp its scale and remoteness from the public highway (SR 240) (Hanford Communities, 2013). From a vantage point to the northwest near where the road crosses the river can be seen what remains of a homestead and barn occupied by pioneers until they were dispatched from the land with only a month's notice in 1942. Nearby on the bank of the river is B reactor, the world's first full-scale nuclear reactor, now a national monument and tourist attraction (Figure 2.3). Slightly inland is C reactor clad in silver grey and now decommissioned to remain in safe storage for at least 75 years. Further downstream are the giant K reactors built for plutonium production during the height of the Cold War. Altogether nine reactors line the south bank of the northern stretch of

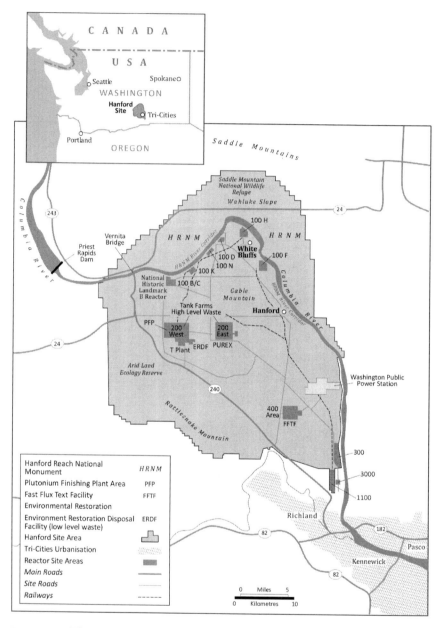

FIGURE 2.3 The Hanford site, Washington State, USA

Map by: John Hunt

the river in the so-called 100 area; all are closed now and in various stages of decommissioning. The B reactor, together with other historic places on the site, now form part of the Manhattan Project National Historic Park comprising sites at Oak Ridge, Tennessee, Los Alamos, New Mexico and Hanford, the three sites which played key roles in the development and production of nuclear bombs during the Second World War. Beyond, the plateau rises with the flat hump of Gable Mountain in the distance. Further south along the highway it is just possible to discern some of the buildings that constitute the 200 area, the core of the site. Here are the vast and sinister 'canyons', the reprocessing works constructed to yield the plutonium from the spent fuel for the bomb. Here, too, are the vast tank farms into which was poured the liquid wastes from reprocessing, some of which have leaked giving rise to fears of seepage towards the Columbia itself. At scattered locations elsewhere on the reservation is a variety of shut down production facilities (fuel manufacturing, plutonium finishing plant, fast breeder test facility), waste management facilities (shallow disposal for low level wastes, waste processing, liquid effluent and encapsulation plants), research and experimental facilities together with the sole surviving commercial nuclear power plant in the Pacific Northwest.

The Hanford nuclear reservation is truly massive in its scale, complexity and significance. In *scale* it covers 586 square miles, about half the size of the state of Rhode Island or, in UK terms, larger than my own county of Bedfordshire (477 square miles) and almost as big as Hertfordshire. The operating area lies within the great bend of the Columbia and it is bordered today to the north and east by the

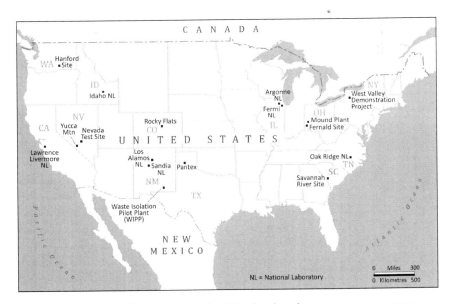

FIGURE 2.4 Major nuclear sites across the US related to the weapons programme

Map by: John Hunt

Hanford Reach National Monument area of protected land and river including the Wahluke Slope beyond the Columbia. To the west lies the almost pristine sagebrush desert designated as the Arid Land Ecology Reserve flanking Rattlesnake Mountain, while to the south lies Richland, the original Hanford nuclear community now part of the wider Tri-Cities. In terms of *complexity* there are some 1200 individual sites where waste or materials are stored, 400 along the Columbia and 60% (700 sites) in the central part of the site (Gephart, 2003). In this part of the site are the 177 tanks holding high level wastes or residues as well as past liquid disposal sites and solid waste burial grounds. The total inventory of wastes and contamination is unknown, perhaps unknowable. Nonetheless, best estimates suggest 350 million curies of radioactivity and 400,000 tons of chemicals now exist in Hanford's tanks, buildings, soil and groundwater.

As to *significance* Hanford is the largest clean-up project anywhere in the world at the present time and receives an annual inflow of around two billion federal dollars for the purpose. Hanford takes around two-fifths of the US Department of Energy's (DOE) clean-up appropriations for its 17 major sites (Figure 2.4). Hanford is a paradox, a place of seemingly inconsistent and conflicting contrasts.

> It contains a portion of the nation's most dangerous radioactive waste while preserving some of the most unique desert ecology within the Pacific Northwest ... Hanford contrasts yesterday's certainty for achieving plutonium production goals with today's uncertainty about how to reach cleanup objectives. Plutonium production was done in secrecy; cleanup is an open book being written through an evolving social consensus and understanding of the problems faced.
>
> *(Gephart, 2003, p.v)*

During its relatively long history, Hanford's development has reflected the shift through the various discourses I described in the previous chapter. It is true that throughout its history Hanford has epitomised the fourth of those discourses, the discourse of Security, for its very essence was the defence of the nation against external threat. But, it is also possible to trace an evolution through the first three ages, too. First, there was the secretive, closed and deferential culture of the war years, then came the aggressive yet defensive, reactive and somewhat introspective though purposeful sense of self-awareness and pride in its national role during the Cold War. Together these phases reflected the discourse of Trust in Technology expressed in the statement by President Kennedy quoted at the head of this chapter. With the publication of once secret documents detailing previously unacknowledged contaminant releases and ending of the Cold War in 1989 came the sense of shock and betrayal with loss of self-confidence and feelings of isolation and bewilderment as Hanford's mission shifted suddenly from plutonium production to environmental clean-up. The turbulence of this provoked a relatively short-lived phase of the discourse of Danger and Distrust. Latterly has emerged a more settled phase with clean-up becoming the generally accepted mission backed by political and economic support and with Hanford gradually relinquishing its overdominance of the local region. This has been a period that reflects the discourse of Consensus and Cooperation, echoing more the values represented in President Johnson's thoughts also quoted at the beginning of this chapter. Although there remains a modicum of residual reluctance to accept that production is finished at Hanford, it seems unlikely that any perturbations, internal or external, will interrupt the generally settled social conditions of the Hanford communities.

Changing discourses – from production to clean-up

The perfect location – a 'wasteland with sagebrush vegetation'

In December 1942 Lieutenant Colonel Franklin T. Matthias was sent on a secret mission by General Leslie Groves, the military head of the Manhattan Project. The Project had assembled a group of atomic physicists who had successfully produced Uranium 235 and Plutonium 239, the two fissionable materials for atom bombs. Groves' task was to realise the military potential of these discoveries by constructing the plants that would produce atomic weapons of devastating power. In the exigencies of wartime it was imperative that such plants could be built quickly and covertly. Oak Ridge in Tennessee was selected for developing U235 for the bomb eventually dropped on Hiroshima, and Los Alamos, New Mexico, was the setting for the laboratories for designing and assembling the bombs. A third site was needed to provide a safe and secure area for the dangerous task of making plutonium. Accordingly, Matthias was ordered to find a location for the production of plutonium sufficiently large and isolated to ensure the separation of plants for safety and with ample supplies of water and electricity for large scale operations. In eastern Washington he found the ideal location. Here, for about 60 miles, the broad Columbia flowed between high bluffs making a long sweeping bend known as the Hanford Reach. Beyond it to the west lay a treeless plateau, interspersed with farms and apple orchards but mostly devoid of human settlements save for Native American tribes that seasonally roamed this area, bordered in the distance by the isolated and foreboding ridge of Rattlesnake Mountain, which provides a prominent and familiar landmark in the region. Both upstream and downstream were vast hydro power projects including the giant Grand Coulee dam which would provide Hanford's substantial but secretive demands for electricity (Ficken, 1998; Lee, 1989). This vast sweep of country perfectly matched the siting criteria and Matthias had no hesitation in recommending it to Groves. 'I thought the site was perfect the first time I saw it', he said (quoted in Sanger, 1989, p.6). Indeed, Matthias concluded, 'the site was so good that there couldn't be a better one in the country. It looked perfect in every respect' (quoted in Findlay and Hevly, 2011, pp.18–19). In its remoteness it seems also a textbook example of a peripheral location.

Over the following two and a half years the pace of development at what came to be known as the Hanford Engineer Works (HEW) was frantic. There was a similar urgency at the other two wartime atomic complexes at Oak Ridge and Los

FIGURE 2.5 Colonel Matthias with Wanapum leaders in 1942

Source: US Department of Energy

Alamos. Groves was driven by three imperatives; the desire to win the war quickly, the fear that the Germans might produce the bomb first and the possibility that the war might end before the project was completed and a bomb had been detonated (Goldberg, 1998). Within three months of the selection of Hanford the land was expropriated; farmers were told to leave their homes and the Native Americans informed they would no longer be able to fish and gather in the area of the Hanford Reach except under strict supervision. In the extraordinary circumstances of wartime there were conflicting reactions among those displaced. Among some there was passive acceptance of loss of home and livelihood (the compensation only covered the value of the land) and 'the farm families and villagers dutifully departed for the good of the war effort' (D'Antonio, 1993, p.14). For other landowners there was resentment at the speed of eviction and the loss of land they had nurtured with some getting better compensation through the courts (Findlay and Hevly, 2011, p.21). The tiny settlements of Hanford and White Bluffs were evacuated. Nothing now remains of White Bluffs except for a historic sign and the derelict grey concrete shell of the town bank.

FIGURE 2.6   White Bluffs High Street before 1943 and remains of bank. Bottom
picture shows surviving stone warehouse of farm from which owners
evicted as part of the Hanford clearance, now part of the Manhattan
Project National Historic Park

Source: Top left: US Department of Energy; top right: author; bottom: Cynthia C. Kelly, courtesy of the
Atomic Heritage Foundation, USA

The Hanford camp became 'the largest such construction settlement ever assembled
in the United States' (ibid., p.58) with, at its peak in 1944, over 50,000 workers
recruited from all over the country and housed in grey wooden barracks with strict
segregation of men and women and whites from blacks and Chicanos,

> They came to Hanford by the thousands, some stayed for the life of the job,
> and beyond into their retirement years. Others didn't last any longer than the
> wait for the next train out of Pasco. Others had no idea what was happening
> at the Hanford Engineer Works, only that something was going on there that
> would win the war.
>
> *(Sanger, 1989, p.47)*

They built three reactors or piles (identified by letters as B, D and F) on the banks
of the Columbia River where they could extract and discharge the large volumes
of water needed by their single pass operating systems. The spent fuel from the
reactors was transferred inland to the central plateau (200 Area) where plutonium

FIGURE 2.7 Aerial photo of Hanford Camp barracks, 1940s

Source: US Department of Energy

was separated out by chemical processes in massive plants known as 'canyons'. (The post-war PUREX plant which produced three-quarters of the plutonium manu-factured at Hanford was 1,000 feet long and 100 feet high.) As Geoff Harvey, the PR Manager for the contractor BNFL, told me on a site visit in 1999, 'back in the 1940s only six to eight feet of concrete stood between you and the most dangerous material in the world'. From these operations wastes were discharged to the atmosphere, ground and, in the case of highly radioactive wastes, to tanks stored on site.

Elsewhere in the southern part of the site (300 Area) a fuel fabrication plant was also constructed along with a variety of research, maintenance and administrative facilities. By the end of the war there were 1,500 buildings and 64 underground high level waste tanks on the site (Gerber, 1993, p.7). To the south of the HEW the town of Richland, a company town reserved for federal employees, had already reached a population of 13,000.

With the detonation of the uranium bomb (code named 'Little Boy') on Hiroshima on 6 August 1945 President Truman revealed the secret of the wartime operations at Hanford and the other sites. 'It's Atomic Bombs' was the headline in *The Villager*, the Richland newspaper, and, following the Nagasaki plutonium bomb ('Fat Man') exploded on 9 August and the subsequent capitulation of Japan, the newspaper announced on 14 August 'Peace, Our Bomb Clinched It'. 'The sphere of plutonium 239 in the Nagasaki bomb weighed 6,100 grams. One gram of plutonium, or one-third the weight of a penny, transformed its mass into pure energy to produce the explosion that destroyed Nagasaki's Urakami valley' (Del Tredici, 1987, p.129).

Although the bombings were justified in terms of bringing the war in the Far East to a swift conclusion it has also been argued that the Japanese were on the

FIGURE 2.8 Hanford's historic T plant where plutonium processing first took place

Source: Author

FIGURE 2.9 'Fat Man', the bomb that was dropped on Nagasaki on 9 August 1945

Source: US Department of Energy

FIGURE 2.10 Headlines in Richland's weekly newspaper, *The Villager*

Source: US Department of Energy

point of capitulation and the terrible massacre of the civilian populations of two large cities was unnecessary and unjustifiable. But it became clear later that General Groves who had been responsible for the successful military prosecution of the Manhattan Project at Oak Ridge, Los Alamos and Hanford was determined to see through the conclusion of his work by unleashing both the uranium and the plutonium bomb with three days in between. As Stanley Goldberg observes,

> It seems clear today that the rush to produce the active materials and to drop the bombs on Japan as soon as possible was driven largely by a fear that the war might end before both types of fission bombs could be used.
>
> *(Goldberg, in Hevly and Findlay, 1998, pp.74–5)*

The moral questions emerged much later. At the time the Hanford community was suffused with pride at its contribution to the war effort. As Michele Gerber, the historian of Hanford, commented, 'Nothing can make you that proud ever again' (interview, March 2004). The process of transformation in mid-Columbia found its expression not simply in the nuclear landscape but in the sense of newfound identity widely felt by an incoming community that had produced the materials for the bomb. In the fury of wartime production the consequences for the people and the environment were scarcely considered but were nonetheless profound. 'You do not bring thousands of people and immense tonnage in materials and equipment onto this stark and lovely desert without changing, perhaps, forever, the delicate balance of life' (Gerber, 1992, p.30). Over the coming years the interaction of land and people created and shaped the changing political geography of the Hanford site and its region.

The years of expansion

For two decades after the war Hanford experienced growth in production and population to provide the plutonium needed for the arms race. This was a period of pride and trust in the technological capability of defending the country and defeating its enemies. The expansion came in phases stimulated by world events, first the onset of the Cold War in 1947 and the explosion of the first Soviet atomic bomb in 1949, then the Korean War in the early 1950s and continuing through the period of international tension of the early 1960s. A further five reactors were built along the Columbia River (H, DR, C and the twin 'jumbo' reactors KE and KW) standing 'like American pyramids along the Columbia River' (Gerber, 2001, p.3). This surge in production capacity was matched by an increase in processing facilities to take over from the wartime reprocessing canyons (T and B plants). The new facilities were the reprocessing plants, REDOX (reduction oxidation) plant commissioned in 1952 and the PUREX (plutonium uranium extraction) plant in 1956 which fed the Plutonium Finishing Plant (PFP) and the Uranium Trioxide Plant (UO_3) (Gerber, 1993). These reached peak production by the 1960s. All this required waste processing facilities including recovery plants and a further 85 storage tanks. By the early 1960s the workforce on the site was nearly 10,000 and its economic base had been sustained over many years.

From the outside Hanford seemed to be 'one of those secret places, where no-one wanted to live' (Dan Metlay, senior official with the U.S. Nuclear Waste Technical Review Board (NWTRB), interview, March 2004). In the early years the activity of Hanford was so secret that even employees did not know what was the purpose of their work beyond a belief in its necessity in the national interest. Susan Gordon of the Alliance for Nuclear Accountability explained to me how the components of the nuclear complex were scattered across the country thereby erecting further barriers to comprehension of the nuclear project as a whole (interview, 2004). The public image of the sites was often misleading or downright deceptive. Thus the Fernald Feed Plant in Ohio suggested a food factory whereas in reality it was a uranium refinery. The Pantex plant in Texas which manufactured warheads was sometimes mistaken as an electronics factory. And, even at Hanford, the PUREX plant suggested the manufacture of bleach rather than plutonium (D'Antonio, 1993, p.236).

Viewed from within, Hanford's communities were prospering and developing rapidly. The increase in production was matched by the development of the urban area. By 1962 the population of the Tri-Cities was 54,000 with around half in Richland, the company town just to the south of the reservation. In its early years Richland resembled a 'closed city' according to former mayor Larry Haler (interview, October 1999). Until 1958, when the city was incorporated, it was controlled by the government. It is said that even the light bulbs had to be changed by those specially assigned for such tasks. Only those working at Hanford were housed there and they were required to leave their homes on retirement. Richland's population was well educated (it was frequently claimed that the city had more PhDs per head than any other city in the United States), well paid, youthful (with the highest birth rate in the nation in 1948) and sociable, supporting a myriad of clubs and sports facilities totalling around 250 voluntary associations by the 1980s. It was somewhat homogeneous socially compared to other cities in the country, an egalitarian community but also hierarchical as evidenced by the juxtaposition of housing types differentiated by jobs and income. The early 'ABC' or alphabet pre-fabricated housing (so-called because house types were designated by letters of the alphabet) was constructed of good quality wood in generous mixed layouts amid ample green spaces (Columbia River Exhibition, undated). The houses designed for the so-called 'upper echelon' of top administrators, scientists and physicians were detached with three or four bedrooms and a monthly rent of $67.50 in 1949. There were mid level (managers and engineers) smaller detached or semi-detached three bedroom homes renting from $33.50 to $50.00. Blue collar homes rented at $25.00 to $35.00 depending on size and there were also single-sex dormitory blocks. The design and layout of Richland was redolent of the ideas of social engineering through the socially integrated neighbourhood unit concept inspired by Clarence Perry and influential in post-war planning schemes in Britain and the USA (Perry, 1929). It had a generous layout organised in neighbourhoods of around 5,000 people with open space, shopping malls, schools and public buildings. Richland's planning reflected the communitarian ideals of the early garden city movement though later, as it moved from public to private ownership by the 1960s, it took on more and more the suburban planning vocabulary that typified urban design in an age of greater prosperity (Abbott, 1998). But, in the early years, Richland seemed an expression of the ideals of community and participation which emerged from the shared and patriotic experience of contributing to the war effort.

The spirit of unconditional support for a project which, though secret, was regarded as vital to the country's defence is evoked in the reminiscences of the early workers at the site. Lee Burger who came to work in the solvent extraction

FIGURE 2.11   'Alphabet houses', Richland

Source: US Department of Energy

FIGURE 2.12   Atomic Frontier Days annual celebration of 'new light on the old frontier'

Source: US Department of Energy

plant in 1948 commented, 'Most of us didn't know what was going on', but nevertheless the universal attitude was that 'This is important – it's something we should do' (interview, March 2004). Even then there was a sense of being outcasts. 'There was a tendency to look down on this place. A lot of people didn't want anything to do with it'. This combination of pride in the place and its work and defensiveness at its lack of recognition is evident in the commentary of Wanda Munn, an engineer who came to work on the Fast Flux Test Facility (FFTF), a neutron research reactor for testing the commercial design and operation of breeder reactors. 'The community really has been a company town. We loved the company because it defended our country' ... 'Nuclear has had a bad image but it's clearly the fuel of the future, the greatest hope for least environmental damage. It's astonishing technology operated by scientists and technologists of the highest calibre' (interview, October 1999). Brian Freer in his study of the first generation of workers at the Hanford site identified a combination of positive attitudes characterised by a faith in technology, a sense of purpose, a strong sense of belonging and an attachment to the area and its 'history' (Freer, 1994). Although there were, even in those early years, concerns about stigma and the awesome destructive power of Hanford's production, these feelings tended to be muted in a condition of willingly repressed awareness. Hanford's defence lay in its purpose in the defence of the nation.

Hanford's expansion had been defined and justified by a discourse of the needs for national defence against the dangers of external aggression. Later came the idea that Hanford was contributing to peace both through its production of weapons and its potential for the peaceful uses of atomic science. The N reactor, designed to produce the material for weapons grade plutonium as well as electricity, and this dual purpose was reflected in President Kennedy's speech quoted at the beginning of this chapter. But, by the time such projects were under development, Hanford had already begun its decline as a national defence production site.

The years of transition – from plutonium culture to environmental culture

By the mid-1960s the Hanford site, or reservation as it became known, was beginning a long period of transition which lasted for around three decades. Up to this point Hanford's military mission had been unambiguously clear, even if the details were largely kept secret. A distinctive culture had suffused the community which Gephart (2003) calls a 'plutonium culture'. Although he recognises that cultures are complex, dynamic interactions of values, beliefs and assumptions it is possible to discern a dominant culture reflecting the discourse of Trust in Technology prevalent throughout this period. 'At Hanford, trust in officials and secrecy underpinned the early years' (p.2.17). This was sustained so long as the local community's undeviating belief in its mission was in accord with the wider perceived national interest. But, as the arms race gradually receded to be succeeded by the prospect of arms reduction and then, at the beginning of the 1990s, by the collapse of the Soviet Union so Hanford's role in national defence gradually diminished. President Lyndon Johnson announced the impending closure of the reactors in the mid-1960s and the Nixon–Brezhnev era of détente began in the 1970s with the signing of the first weapons limitation treaties.

The decline in military production can be measured in terms of fuel reprocessed at the site. This had reached around 7,000 metric tons per year in the early 1960s and, after reaching its peak in 1965, thereafter declined, partly as a result of more efficient processing when the N reactor replaced the single pass reactors but also in response to reduced demand. From 1972 production at the remaining reprocessing (PUREX) plant ceased for 11 years and, after a brief revival with much lower output, it ceased operating altogether in 1990. Employment on the Hanford site had fallen below 7,000 in 1971.

The temporary shut down of the N reactor (following Chernobyl) in 1987 followed by its permanent closure finally brought to an end Hanford's military mission. For a time it seemed Hanford's future might lie in nuclear research and development of peaceful uses of nuclear technology. Among the non-military projects were the Fast Flux Test Facility (FFTF), a breeder reactor that only operated for ten years and the nuclear power plant operated by the Washington Public Power Supply System which was devoted entirely to electricity production. The closure of the FFTF, which occurred in 1991 and which was confirmed in 2000, provoked a final conflict between supporters of a continuing production at Hanford and opponents who wished to see it focused entirely on clean-up. The decision effectively ended the prospects of a future based in new production technology. But there is still the glimpse of some future diversification not unconnected with its former role. The Mid-Columbia Energy Initiative has a series of potential missions and investment in an array of energy projects including the possibility of small modular reactors which might come to Hanford Energy Parks under the Mid-Columbia Energy Initiative (Gary Petersen, interview 2013).

For a time Hanford appeared to have a possible future as the disposal site for the nation's spent fuel from its civil nuclear reactors. Under the Nuclear Waste Policy Act (NWPA) of 1982 two sites for deep geological repositories, one in the west and one in the east of the country (soon dropped as a result of widespread opposition in the eastern states), would be nominated. Already, in the west, nine sites, including Hanford, were under investigation and the list was subsequently reduced to five, Hanford being ranked fifth in terms of suitability. However, when the final three sites were nominated, Hanford's political support promoted it to second place on the list, after Yucca Mountain, Nevada, and ahead of Deaf Smith, Texas. The choice of sites officially rested on the need for comparison of diverse geologies and Hanford was the only site in basalt. Although it was not regarded as the best site geologically, Hanford possessed potential political advantages as the one area where local (as opposed to state) support could be anticipated. However, the high costs of investigating three sites before reaching a decision, together with the political possibility that the three states (Nevada, Washington and Texas) could combine to oppose the project, led to the NWPA Amendment in 1987. By this time Yucca Mountain was selected as the only site for characterisation i.e. detailed study of suitability, thereby isolating Nevada to the relief of other states (Blowers et al., 1991). Over the next three decades Yucca Mountain was sufficiently characterised and a licence application was made in 2008, at which point the Obama administration suspended further work on the project. That is another story but Yucca's potential demise is unlikely to resuscitate Hanford as a host for the nation's wastes. It is geologically unsuitable and has quite enough problems managing the complex brew of wastes already on site.

During the years of transition as Hanford's role changed so, too, did knowledge and understanding of the reservation's relationship to the community. Pride in achievement and in the potential of nuclear science remained strong but it became mixed with apprehension about the economic future and the health and environmental risks associated with the site. With its original mission gone and the possibilities of new production roles stillborn it was becoming clear that Hanford's future lay predominantly in the clean-up of its legacy. This required a painful adjustment in attitudes and beliefs about Hanford as the workers and community faced up to the fact that Hanford's role in the defence of the nation had been replaced by its role in defence of the environment. Gradually a plutonium culture gave way to an emerging environmental culture.

Clean-up, Hanford's new role

> Under the plutonium culture, few environmental problems were openly acknowledged. Under the new culture, problems suddenly abounded, and openness became more than an option; it was a prerequisite for economic survival.
>
> *(Gephart, 2003, p.2.17)*

The recognition of its changed role was like 'a birthing process' according to Eric Olds of the DOE (interview, March 2004). It gradually became apparent how serious the problems were in 'the little-known reservation that is arguably the most polluted place in the western world' (D'Antonio, 1993, p.1). From its inception in 1944 until the plants finally shut during the late 1980s, 'Hanford released radioactivity intentionally and routinely during its operations' (Findlay and Hevly, 2011, p.58). From the reactors and reprocessing plants waste products have poured out into the river, the atmosphere and the ground. Water from the early single pass reactors raised the temperature of the river water and, despite some retention, wastes containing radionuclides were discharged into the Columbia posing threats to salmon fishing and drinking water supplies. The shut down of the reactors removed the direct impact but left a major problem of waste storage trenches and contaminated soil near the Columbia with the danger of seepage of complex waste streams into the river.

The atmospheric contamination was even more widespread from windblown contamination of soils and their potential harm to crops and animals as well as humans. Yet the culture of secrecy withheld revelations of the hazards until 1986 when the then site manager Mike Lawrence decided to publish the records in a series of documents amounting to 19,000 pages. Lawrence believed that 'what went on here was good and necessary' to win the war and ensure the nuclear deterrent during the Cold War. But 'it was very secretive, we know best. The secrecy and arrogance has hurt the industry. How can people understand if they are not told?' (interview, October 1999). By revealing the records he intended to usher in a new era of openness which would dispel the mystique surrounding the plant. It was a turning point in the shifting of discourses and their attendant cultural responses.

What the records did reveal was the sheer scale of the legacy and the casual attitudes to risk that had prevailed in the past. This had been a time when reactors and plutonium reprocessing came first and second and the environment a rather poor third in priorities. One of the most serious incidents was the notorious 'Green Run', which took place in 1949. This was a deliberate experimental release of over 7,000 curies of iodine-131 and other radionuclides from raw or green uranium subjected to a much shorter cooling period than usual and which, consequently, had a higher activity and emitted more pollution. The plume of radioactivity extended 200 by 40 miles north east and south west of Hanford giving readings which exceeded the contemporary exposure standards by 11,000 times on the reservation and hundreds of times in downwind communities. It occurred in the wake of the first Soviet nuclear tests. It later emerged that the Green Run was an experiment to simulate the production capabilities of Soviet plutonium production plants in order to develop a monitoring methodology that would enable the US to estimate Soviet bomb making capacity. As Jerry Gough, a nuclear historian, commented, 'The atrocity of the Green Run was not the release itself but the fact that they didn't know what its effects might be. This was outrageous' (interview, October 1999).[1]

During the 1980s the revelations exposed Hanford to both local and national attention. Michael D'Antonio in *Atomic Harvest* (1993) describes how local farmers like Tom Baillie raised concerns about possible impacts on health from airborne contamination. He records how safety inspections by Casey Ruud and others revealed failures of maintenance and management which could increase risks to workers and the environment. These and other incidents were constantly reported in the press (notably by reporter Karen Dorn Steele in the Spokane *Spokesman Review*) and helped to shift the culture surrounding Hanford from one of confidence supported by secrecy to one of defensiveness in response to exposure and greater openness.

Teri Hein in *Atomic Farmgirl* records growing up in the rich rolling wheatlands of the Palouse downwind of Hanford on the Washington/Idaho border. She reveals how an age of innocence gradually shifted to a time of suspicion as the incidence of cancers and other diseases were increasingly attributed to the secretive operations of the plants over 50 miles away. Even then in this introspective rural world anger was mingled with an initial patriotic reluctance eventually becoming a resentment against the unknowable hazards inflicted on the area.

It's hard to look out over those fields ... and consider the hills as poisoned ... If it turns out that we all got poisoned for the war effort, we can know that it is more than just enlisted people who were sacrificed.

(Hein, 2003, p.237)

As production ceased the problems of air and river pollution decreased. But, the leaching of radionuclides into the soil and groundwater, with contamination eventually reaching the Columbia about 20 years after reprocessing operations began, are the enduring and most intractable legacies at Hanford. During the production period the low level liquid wastes from the 200 Area operations were siphoned off into cribs and swamps. Efforts were made to manage the flows of low level discharges to 10% of concentration guidelines but they nonetheless made their way to the groundwater in higher concentrations. High level liquid wastes from the 200 Area were discharged into tanks. These HL tanks are reinforced, concrete lined with carbon steel and their dome shaped tops are 6 to 8 feet below the surface. The largest tanks are 75 feet each in diameter. Tank volumes range from 55,000 gallons to nearly 1.2 million gallons. There are 177 tanks altogether (149 single shelled and 28 double shelled) containing about 55 million gallons of wastes deployed in tank farms in the 200 Area. The single-shell tanks were built to last around 25 years but some leakage is known to have first occurred by 1959. It was established that 67 of the older single-shell tanks are known or suspected to have leaked up to a million gallons of waste and pose a threat to the underlying

FIGURE 2.13 High-level radioactive waste storage tanks under construction at Hanford, c. 1943

Source: US Department of Energy

groundwater which is moving slowly to the Columbia (Zorpette, 1996; USDOE, 2013). The plan was to pump the liquid waste from the single-shell to the double-shell tanks leaving the residual and intractable saltcake and sludge to be dealt with either by condensing, leaving in situ or through other means. Ultimately, most of the tank waste was due for vitrification. Although pumpable liquids were removed from the single-shells by 2004, further complications were encountered and by 2013 leakage was first detected in one of the double-shelled tanks. As of 2013, five additional single-shell tanks were identified or suspected to be leaking waste into the subsurface. The management of these tank wastes is the single biggest challenge faced in Hanford's clean-up, they are the 'watershed issue' (Eric Olds, interview, March 2004). Strategies have changed many times and vitrification may not be the only way. 'Dealing with the tanks has become a perfect monster' (Bill Dixon, interview, May 2013).

It is almost impossible, indeed futile, to try to estimate the overall levels of contamination and inventory of radioactivity at Hanford. For one thing records were not kept in the early years and some that were kept were destroyed. In any case the levels vary according to the levels of production and the half-lives of the different radionuclides. The potential health and environmental impacts are almost impossible to estimate given the intervening variables that affect both the dose and response to radioactivity. But what we do know is that Hanford has by far the largest volume of high level wastes in the DOE's defence sites arising from its complex reprocessing operations, together with 11 tons of plutonium, solid wastes contained or dumped and a vast number of contaminated facilities scattered around the reservation. One estimate, provided for my 2004 visit, suggested there were 1,700 waste sites and 500 facilities to be decommissioned or decontaminated including the nine reactors being put in safe storage, and four processing plants. It estimated 450 billion gallons of liquids had been discharged to the soil with 5 million cubic yards of contaminated soil and 80 square miles (270 billion gallons) of groundwater contaminated above drinking water standards (UK Nirex, 2004). But, 'there are many uncertainties and unknowns' and 'deciphering this entire inventory is less important than pinpointing, or at least bounding, those portions posing the greatest potential health risk' (Gephart, 2003, p.5.3). This is at the heart of the clean-up process and dealing with environmental and health threats eventually became the fount of the developing discourse of Consensus and Cooperation. The transition was marked by a brief though recognisable period of discursive instability as the tenacious grip of the formerly dominant nuclear culture gradually gave way to the emerging focus on clean-up.

Change and continuity in Hanford

This transitional period was relatively short covering a period from the late 1980s to the early years of this century. It was a time of coming to terms with the changed circumstances surrounding Hanford, marked by a reluctance on the part of an established community to let go of production and a resistance to its new role and a resulting yearning to grasp any activities that might prolong the nuclear culture. But, this monocultural dependence had long been waning, becoming increasingly defensive in the face of changing demographics and diversity within the region. The singular plutonium culture had given way more and more to the complex overlapping of cultures and contrasting beliefs that reflected a turbulent period of conflict and change.

This changing culture of Hanford over the years is reflected in the responses of the local community and the wider public to its problems. Initially, in the closed and somewhat claustrophobic community of wartime and the early post-war period, there were the symptoms of what Ron Kathren, a health phycisist at Washington State University, and Wanda Munn call the 'Hanford syndrome', a resistance to outside interference, a belief that the government would take care of everything and an attitude that if it's not invented in Hanford then it doesn't count. By the 1960s, this introspective attitude had begun to change, partly as a result of newcomers who did not share the founding beliefs of the local society. There was a shift from the notion of a benign atomic energy towards the idea of a beleaguered DOE site. According to Ron Kathren, the local community was becoming 'confused, inconsistent and unable to understand the strong opposition and difficulties of the site' (interview, 1999). Nevertheless, the Tri-Cities continued to defend the industry on which they had come to depend. As Pam Brown, Executive Director of the Hanford Communities, an organisation promoting the common interests of the Tri-Cities, put it in 2004, 'This is nuclear-friendly country. We want nuclear activities here. If people are afraid of Hanford, they don't live here' (interview) and again in 2013, 'People who came here were comfortable with nuclear or they wouldn't have come here'.

Opposition has tended to be most vociferous from surrounding areas. Richard Morrill characterised Hanford's isolated position within the wider region as 'a classic case of the supremacy of regional and national consumption values over local production values' (1999, p.2). Within the local region the nuclear reservation and farming have coexisted creating a tension between 'the benefits of the atom and the bounty of the land' (D'Antonio, 1993, p.87). The concerns about contamination of soil and impacts on health of the 'Downwinders', farmers and small town and rural communities surrounding the Tri-Cities, have gained wide publicity and generated suspicion about the activities within the reservation. They have certainly contributed to the culture of greater openness and emphasis on the need for clean-up. But, in the Tri-Cities themselves, support for Hanford remained strong. Opposition from farther away, from the cities in the west of the state, from Spokane and from Oregon, stimulated feelings of resentment and fears of subordination to the interests of more populous and powerful areas. These feelings inspired the formation in 1986 of the Hanford Family, a pro-nuclear group whose purpose was to promote Hanford and defend its contribution to the area. The Hanford Family staged a series of rallies and meetings and sustained a highly visible and articulate opposition to environmental groups, notably Hanford Education Action League (HEAL) based in Spokane. The Hanford Family confronted its opponents on its own territory in John Dam Plaza in the heart of downtown Richland in 1987 shortly after the N reactor was shut down. The Hanford Family campaigned for the restart of the N reactor which was the lynchpin on which all the production facilities in the complex depended.

When it became evident that the N reactor would not be reopened and that Hanford no longer had any role in nuclear production, the tensions within the community began to abate. Defenders of Hanford now had to turn their attention to securing its future role and jobs in clean-up. With its plants now closed Hanford's opponents no longer had targets in their sights. They, too, could pitch their energies into promoting the clean-up mission. A new and more cooperative phase (redolent of a discourse of Consensus and Cooperation) began to develop though residual conflicts remained over the timing, cost, pace and objectives of clean-up. Relationships between local, state and federal interests became more tolerant as Hanford began to adopt and define its new role.

Hanford's history is represented in its political geography. The nuclear complex overwhelmed the pre-existing rural agricultural economy although Pasco maintained a role as a regional agricultural processing and distribution centre. The nuclear complex stimulated the rapid growth of the Tri-Cities and their emergence into a sub-regional centre. But, despite the development and diversification, the area is still, in certain respects, peripheral. Once a frontier region in the American West, the nuclear industry sustained a pioneering mentality and Hanford, to this day, retains geographical, political and social characteristics of the periphery. But, its peripheral character shaped in its days as Atom City of the West (Findlay and Hevly, 2011, p.95) has been reformed and transformed as its role and relationships have changed. By the beginning of the century the cultural transition was almost complete and a new, diverse culture had developed. Hanford's role as a centre, perhaps an exemplary centre, for clean-up was firmly established, economic dependence on federal funding was still strong but no longer dominant and, after the uncertainty, conflict and vulnerability that marked the years of transition, a more settled period of stability, prosperity and consensus had taken root in the Tri-Cities region. I now turn to my third theme of exploring the concept of peripheral community using Hanford – in its geographical, economic, political, social and environmental context – to try to understand its role and relationships in the contemporary world.

Hanford – peripheral but powerful?

Geographical remoteness – the reason for survival

As a military reservation Hanford owes its existence to its remoteness. Besides ample water and electricity the site met the other locational criteria based on a fairly primitive knowledge of radiation dangers which were set out by General Groves in 1942. Its size should be around 12 miles by 16, employees should be housed at least 10 miles upwind of the plants, no town should be closer than 20 miles and no main road or railway closer than 10 (Goldberg, 1998, p.50). As soon as he saw it Colonel Matthias had exclaimed, 'This is it' (Pam Brown, interview, May 2013). At the time there were scarcely 300 people in Richland. As Michele Gerber commented to me, the military 'liked the isolation and didn't have to displace large numbers of people' (interview, March, 2004). During the years of expansion the boundaries of the site were extended to a total area of 574 square miles (1,450 sq. km).

Stretching about 30 miles north to south and 20 miles east to west, Hanford is one of the largest DOE military sites in the US (Figure 2.3). Today, the signs of clean-up activity are everywhere. Along the Columbia the reactors are being put into safe storage, spent fuel has been removed from the reactor cores and contaminated soil from the banks of the river. In the centre the canyons and other buildings are being decommissioned. Nearby, the leaking tanks are being remediated and the plutonium in the Plutonium Finishing Plant (PFP) has been stabilised and removed. The rusting hulks of redundant nuclear submarines lie high and dry on the surface so that they are visible to Russian satellite monitoring to ensure compliance with international arms reduction agreements. The Waste Treatment Plant (WTP) originally a $2.3billion project for the vitrification of high level wastes, by far the largest investment on the site, is under construction with costs already rising to over $12billion in 2013. Outside the industrial areas rural land uses prevail. To the west of Highway 240 which runs across the site is the Arid Lands Ecology Reserve leading up to Rattlesnake Mountain. North of the Columbia River is another buffer zone known as the Wahluke Slope and including the Saddle Mountain National Wildlife Reserve. Most of the area remains (in 2013) fenced off and strict security prevents unauthorised access to the site. The size and security reinforce the feeling of remoteness and inaccessibility and Hanford remains one of the most impenetrable places in the United States.

As clean-up progresses so the site will become more accessible. Indeed, already the designation of the National Historic Park has promoted greater public interest and access. The DOE reservation will shrink as land is released for other uses. Land to the south will become available for industrial and urban uses while the areas to the north and west will be released for farming and recreational purposes. The presence of the reservation has provided an ecological benefit in protecting the area from the depredations of extensive agriculture and industrial development. According to Gene Schreckhise, an environmental scientist at the Washington State University Richland, Hanford has done a great job of isolating one of the few surviving refuges for the sagebrush habitat and the various grasses such as bluegrass and bunchgrass and the shrubs, flowers and animals that are unique to the region (interviews, 1999, 2004). But, it is also a fragile ecology, vulnerable to alien species and taking a long time to regenerate. Much of the area is likely to remain protected and parts of the central site where the plants and wastes remain are unlikely ever to revert to public use.

Hanford and the Tri-Cities still seem set apart from the rest of Washington state and the country as a whole although the mysteries surrounding the activities on the site have long since disappeared. This separateness is a matter both of geography and history. The Tri-Cities are 145 miles from Spokane, the nearest large city, and over 200 miles from Seattle and Portland, the major cities of the Pacific Northwest. But, while the area remains separate it is no longer so remote. The Tri-Cities are now a significant service centre for the surrounding region and are connected to other parts of the Northwest by interstate highways, the railway, ocean going barges and a growing airport. And, of course, in common with everywhere else, the area is integrated into the electronic networks of communication. The Tri-Cities, with a population of around a quarter of a million, is Washington state's fourth largest metropolitan area and was the fastest growing in the United States during 2010–11.

Hanford's remoteness brought the nuclear industry to the region in the first place. Although production has ceased, the nuclear reservation remains a major activity though, as Gary Petersen Head of the Tri-City Industrial Development Council (TRIDEC) observes, no longer the driving force it once was (interview, May 2013). Despite its changing function and its relative geographical isolation Hanford exerts an influence through its position in the nation's history and in its contemporary role as a major clean-up site. Remoteness, the reason for its existence, still persists in certain respects. Hanford is far from the centre of power. Michele Gerber expressed the disadvantage: 'Here we're in a corner of the country – you can't get further from Washington D.C. The eastern sites are close and get more attention' (interview, May 2013). But, over the years as the area has developed, its population has increased and it has become more integrated with its surrounding region, more accessible and more diversified.

Economic marginality – the challenge of diversification

From the moment when construction began in the war the local economy has been dependent on the fortunes of the Hanford reservation. During its years of production Hanford's economy was, indeed, vulnerable. Jim Watts, a long time trade union leader in Richland, expressed this sense of dependence dramatically, 'If Hanford sneezes everybody around here gets pneumonia' (interview, 1999). And yet, the economy has proved remarkably resilient. Another veteran community leader, the late Sam Volpentest, former Head of TRIDEC, commented that whenever Hanford shed jobs, 'We thought doomsday was coming but it never happened' (interview, 1999). These comments encapsulate the economic contradictions that long seemed to surround Hanford, the oscillation between pessimistic views of economic disaster following the decline of employment on the reservation and the more comforting belief in a stable future based on diversification. As the commitment to clean-up has developed and the non-nuclear economy has begun to flourish, the Hanford Communities have achieved a more secure position and fears of decline have diminished. As Harold Heacock, also of TRIDEC put it, 'As long as things keep working you put off the day of reckoning' (interview, 2004).

Certainly in the early years, apart from a downturn immediately after the war, employment on the site grew in response to the international situation. By the 1960s it had reached almost 10,000, about 40% of the total workforce of Benton county in which Hanford lies (Gerber, 1993, p.17). Then came a downturn to around 7,000 by 1970 followed by another cycle of boom and bust in the 1970s/1980s, reflecting the varying fortunes of the nuclear power plants, the waste disposal project and the N reactor, with a recovery in the 1990s as the clean-up began to get under way. In the early years of the new century employment was holding at around 12,000 working on the site and a further 4,000 or so in various laboratories attracted to the area and supported from federal funds, bringing the Hanford total to about 16,000.

By the end of the first decade of this century it seemed the Tri-Cities had established a sustainable economy. By this time Hanford accounted for around a quarter of the economy while there was vigorous growth in other sectors, notably research laboratories, health services, food processing and in the nearby region the agricultural sector was prospering, especially in wineries and potato production (the biggest source of French fries in the country). The Tri-Cities had begun to achieve regional and even national economic significance (Fowler and Scott, 2009). With its higher than average income levels, unemployment around the national average (8.5% in 2013, US 7.5%, Washington state 7.0%), economic diversification and its sunny climate making it increasingly attractive as a retirement and leisure area, the Tri-Cities manifest an air of prosperity. The monocultural dependence that is the typical condition of the periphery has gone.

Nevertheless, Hanford makes an important contribution to economic stability and, given its $2 billion budget, it is likely to continue to do so. Dependence on federal funds suggests vulnerability. In the early years of the century there was considerable pessimism about the longer term. Harold Heacock, speaking in 2004, believed that clean-up, though it sustains employment for a while, also spells the eventual demise of Hanford as a significant employer. This reflected a general perception at the time that Hanford was going nowhere. A decade later there were good reasons for taking a more optimistic perspective on Hanford's economic prospects, at least in the medium term.

In the first place clean-up is a complex, time-consuming process where costs are difficult to control. There has been enormous effort to deal with the reactor sites along the Columbia River and to move on to the problems at the centre of the site. But, as Roy Gephart points out, there has been a tendency to deal with the easy stuff first. In the early stages it was more basic 'muck and truck' technology 'clearing up the low bearing fruit' such as demolition of buildings and disposing of slightly contaminated soil in a central landfill. By 2013, after nearly a quarter of a century of active clean-up, very little of the site's massive burden of radioactivity had been removed or remediated (interview, 2013). But, even dealing with spent fuel sludges proved horrendously difficult, dangerous and labour intensive as Gerber describes:

Nuclear workers stood on grating above the basins, working through slots with long handled tools to grapple and manipulate each fuel assembly or piece thereof … Often the workers had to toil in heavy, bulky radiation protection suits and breathe through air packs and hoses.

(Gerber, 2007, p.245)

Clean-up of the central area, and especially the tanks, presents more formidable problems. The Waste Treatment Plant (WTP) intended for vitrification of the liquid high level wastes is well over budget and well behind schedule and plagued with problems. It ensures continuing investment without guaranteed success. Overall the clean-up programme is proving far more intractable and durable than was ever envisaged.

A second reason for the protracted process is the way it is managed. According to Bill Dixon, 'The approach has been for the gold standard which makes WTP expensive and long term' (interview, 2013). The reliance on big contracting firms such as CH2MHill or Bechtel on contracts that are frequently changing and often short term also slows things down. Bill Dixon is strongly critical of the process. 'It's halfway between a shambles and working well, though getting better. Clean-up is expensive, the bureaucracy stifling, productivity too low with regulations that are not effective or inappropriate and unions are a barrier'. He points to the low productivity, the changing strategies, the bureaucracy, the lack of competition with few players as factors impeding progress. But, the procrastination sustains a continuing programme which 'should be 30 years and is more likely to be 50 years'.

The third reason for optimism about Hanford's medium term future prospects is the commitment to clean-up. There is a sense of recognition of Hanford's past role which goes way beyond the Tri-Cities and which converts into a sense of obligation represented in the consistent funding of the clean-up. This recognition and obligation is symbolised in the designation of the Hanford Reach National Monument and the historic park site around the original B reactor. More than that, cleaning up Hanford and protecting the Columbia River has been a call on federal resources which has consistently been met. Federal support has been maintained while overall dependence of the economy on Hanford has been declining. The fear of eventual economic decline, so prevalent at the turn of the century, has diminished as clean-up has become more protracted and Hanford's dominance has waned in the face of diversification.

The fear has not altogether disappeared. There are some fears that inefficiency, complacency and sluggishness may eventually prove counterproductive. Some employers are critical of the Hanford workforce. On the other hand, their technical skills and ability are regarded as one of the economic assets of the area. Eventually, as clean-up proceeds there will be a steady and continuing decline in Hanford site-based jobs (apart from the laboratories which may experience some growth) until most of the clean-up work is complete. That time is some way off, at least until the middle of the century, and DOE expenditure levels on clean-up will be sustained and Hanford will continue to make a significant contribution to the regional economy (Greenberg et al., 2003).

Among the community leaders there is a qualified optimism about the future economic prospects. Pam Brown of Hanford Communities, speaking in 2004, recognised the area had a problem, 'But we live here, we have to deal with it. It is manageable'. She emphasised how 'amazingly well' Hanford had done in terms of federal clean-up money and stressed the potential spin-off from the area especially from the national laboratories. Almost a decade later, she perceived a future for the area in clean energy and clean-up with diversification into agriculture, wine and export activities (interviews 2004 and 2013). TRIDEC and other economic development agencies such as the Port of Benton have been active in trying to attract new enterprises and the DOE and the major contractors on the site have invested in a concerted attempt to diversify the economy. Industrial parks have been created, notably a substantial area immediately to the south of the reservation. Optimists point to the success of the Columbia Center, the major sub-regional retail centre in Kennewick and the presence of the medical technology and diagnostics industry as evidence of prosperity and potential for high-tech growth in the area.

Inevitably and inexorably an economic transformation is taking place. The culture has most definitely changed. Hanford no longer seems almost pathetically dependent on state support, a vulnerable and marginal place whose fortunes are determined by decisions elsewhere, notably in Washington D.C. But, Hanford continues to exert a powerful claim on the national conscience. As Lake Barrett, a nuclear industry consultant, explained, the reasons are ultimately psychological. There is the dread factor, a perception of potentially catastrophic hazards from nuclear sites that ensures governments will respond to the demand for clean-up. But there is also the fact of Hanford's role in American history. 'You owe the people of Hanford. You hate to see them hurt' (interviewed in Washington D.C., 2004). He points out the seeming paradox that many mining, steel and textile towns across the United States, though scarred by environmental degradation and toxic wastes, have simply been left to die once their industry shuts down. But, such places have lacked the charismatic identity nurtured from within and recognised from without which has secured Hanford's survival in changing circumstances.

The newfound economic stability of the Tri-Cities, in which Hanford plays a diminished but still prominent part, has induced a culture of institutional inertia in the clean-up process. According to Roy Gephart this is characterised by a lack of focus where the goals are increasingly unclear and social and managerial risks are avoided. 'There is a resistance to things that buck the status quo, a self-protective environment which inhibits change'. However, Hanford is no longer a marginal or monocultural economy, rather it is a vital component of a system which survives because it must be maintained. Hanford is now part of a more vibrant Tri-Cities region and the relationship between the nuclear site and the wider economy is increasingly a comfortable one. Hanford's changing economic role is reflected in its changing political role.

Political position – persistence of power in a changing world

One of the striking features of Hanford's history is its political sustainability in the face of changing power relations. From a position of relative powerlessness, especially when production at the site was running down, Hanford has emerged in an era of clean-up with substantial political leverage. Throughout its early years the significance of Hanford's role in national security was unquestioned and its political influence thereby assured. The interests of Hanford and of the federal government were coincident. During the years of transition its military mission began to diminish and concerns about its impact on human health and the environment made the nuclear complex vulnerable to its opponents. For a time, notably during the 1980s and early 1990s as the once secretive operations were opened up to public scrutiny, Hanford became a focus of conflict about its future role. The power relations became more complex as the DOE began to reassess its commitments to its military nuclear complex (USDOE, 1995) and opposition from the surrounding region began to exert an impact on Hanford's future role. The closure of the N reactor and the failure to resuscitate the FFTF signalled the end of Hanford's role in production and its future engagement in cleaning up the legacy of wastes. But, the goal of environmental remediation and improvement brought together all the various interests in a broad alignment. An era of consensus and cooperation dawned though conflicts over the pace and priorities of clean-up continue. Nonetheless the continuing political commitment to Hanford depends on a set of power relations which, though persistent, are not necessarily permanent.

Hanford's defining political relationship is with the federal government, or more specifically with its powerful agency the DOE. It is this relationship that makes Hanford at once both powerful and vulnerable. It is powerful in the sense that it has always been able to exert influence on the national imagination and its conscience. There was the deeply held belief both within Hanford and elsewhere that, as former site manager Mike Lawrence puts it, 'what went on here was good and necessary' (interview, 1999). The secrecy surrounding its wartime origins and subsequent post-war role in the arms race encouraged a pervasive culture of routine withholding of information and occasional cover up. But, as Lawrence also recognised, the secrecy and arrogance of the early years was potentially damaging which was why, in 1986, he inaugurated an era of openness in order to engender greater trust. In this respect Hanford was ahead of its time, the first DOE site to introduce the kind of openness which became pervasive during the 1990s. It was Hazel O'Leary, then Secretary of Energy, who, responding to growing anxieties about the risks associated with the nuclear military complex, launched the 'Openness Initiative' applying to all DOE military sites in 1993 and announced a 'moral obligation' to clean-up the nuclear legacy. 'In the old days we decided, announced and then defended policy. In the new days we must engage the public, debate, decide, announce and then go forward' (quoted in USDOE, 1995, p.83). According to Kinsella (2001) the initiative presents a 'reassuring storyline' of a department that is 'knowledgeable, experienced, responsible, and technically proficient – equipped with the expertise and vision demanded by the task ahead' (p.180). This sense of obligation to communities that defended the nation is a powerful theme in the DOE. As Paul Dickman, a senior policy adviser in Washington D.C., put it, 'It's difficult to walk away and leave them comfortable' (interview, 2004). It is a remarkable testament to the power exerted by Hanford and the other sites that the DOE's funding has been maintained regardless of the politics of successive federal administrations. Politically speaking, Hanford is not simply an environmental issue; it is a moral issue.

Nonetheless there is wariness also present in the relationship with the DOE (in particular at the federal level rather than with the local field officials). Although the budget has remained firm, there remain concerns about the longer term, particularly once the major problems (the tanks, spent fuel vitrification) have been tackled. There is suspicion that the DOE's aim for accelerated clean-up and for a 'risk based' approach to individual projects are ways of making sure less money is spent in a shorter timescale for a lower standard of remediation. The state of Washington, supported by environmentalists, insists on tough standards of clean-up. But, the DOE resists this on the grounds that it is seeking value for money. According to Paul Dickman, Senior Technical Policy Adviser at the USDOE in Washington D.C., 'a lot of money is being spent to mitigate risk that is very small' (interview, 2013). Without a risk based set of standards there is an open-ended commitment. More cynical is the view that Hanford is 'a money sink where nothing gets accomplished. It's like paying people to baby sit' (Michele Gerber, interview, 2013).

Hanford's ability to exert leverage on the DOE may well diminish in the future as the national political context in which it is situated changes. Hitherto Hanford has been one of 17 major facilities which make up the DOE's nuclear weapons complex. Some of these, like Hanford, are large reservations including the Nevada Test Site, the Savannah River Site (SRS) in South Carolina, the Oak Ridge Reservation in Tennessee and the Idaho National Engineering Laboratory. There are also sites where materials were assembled such as the now decommissioned Rocky Flats (Colorado, nuclear triggers), the Pantex (Texas bomb assembly) and the research laboratories at Los Alamos (New Mexico) and Lawrence Livermore Laboratory (California) (see Figure 2.4 on page 29) (Kuletz, 1998; Gusterson, 1998). Together these facilities make up a powerful combination from several states and, in the past, have nurtured powerful Congressional delegations to plead their cause. Pragmatically speaking, as Brian Costner, an anti-nuclear activist based in Seattle, points out, the bill for environmental management of $6 billion gets lost in the total appropriation within a total defence budget of $300 billion (interview, 1999).

As clean-up proceeds the number of sites will decline and the investment, and with it the collective power, of the remaining sites, will diminish. Already Rocky Flats and the Fernald uranium processing plant in Ohio have been cleaned up and clean-up has advanced elsewhere. There is the worry that support for clean-up at Hanford may diminish as it becomes more politically isolated (Oregon Department of Energy, 2004). Hanford takes the largest appropriation of clean-up funds but it has lost out to some other sites, notably Savannah River Plant (SRP) and Oak Ridge where some production of nuclear materials continues, for example the troublesome and delayed Mixed Oxide Plant (MOX) at SRP intended to use up plutonium as nuclear fuel to achieve the USA's obligations to non-proliferation of nuclear materials. One view, expressed by the late president of TRIDEC, Sam Volpentest, considers that Hanford has a diminishing impact on politics. It no longer attracts the headlines, its role as a site specifically for clean-up is no longer contested and, as clean-up proceeds, it may begin to loosen its grip on the national consciousness. As the conflicts over Hanford become more subdued so it may lose its ability to attract attention and, consequently, the level of federal support to which it has become accustomed. Against this is the view that Hanford will remain powerful, supported by a strong Congressional delegation and the necessity of maintaining the clean-up programme.

Within its regional context Hanford has also experienced changing power relations. The concerns of the Downwinders about health impacts from airborne pollution introduced doubts and anxieties about the activities on the reservation (D'Antonio, 1993; Hein, 2003). Investigative journalists and environmental organisations took an increasing interest which fed upon the release of information about past activities provided by the DOE. Whistleblowers were encouraged to speak out without fear of reprisals and several did so (interview with Tom Carpenter of Hanford Challenge, 2013). Yet the opposition to Hanford's activities barely invaded the core area around the Tri-Cities. During the 1970s and 1980s anti-nuclear protest in the state of Washington focused mainly on the grandiose plans of the Washington Public Power Supply System (WPPSS) to develop five nuclear power plants, three of them to be situated on the Hanford site. Protest focused on the plants nearest Seattle. Distant from the West Coast and pro-nuclear, Hanford had been identified as 'a difficult target for protest' and 'local opposition there was virtually non-existent' (Pope, 1998, p.239). In the event, WPPSS suffered a financial implosion and only one plant, at Hanford, survived to produce electricity for the public supply system.

Although there is little visible protest in the Tri-Cities area, Gerry Pollet of Heart of America Northwest detects a growing willingness of people to speak up about health and safety concerns (interviewed 2004). The more vocal opponents tended to come from further away, from Spokane where HEAL (which closed in 1999) was based, or beyond the 'Cascade Curtain' in the populous western part of the state around Seattle and Tacoma and in neighbouring Oregon, centres for the so-called new social movements which emerged during the 1970s and after (Pope, 1998). Here NGOs like Heart of America Northwest (focused on Seattle with few of its members in the Tri-Cities), Hanford Challenge and Hanford Watch (Portland) are based and take a close interest in Hanford with seats on the Hanford Advisory Board, a consultative body comprised of public and stakeholder representatives.

Until recently, Hanford provided a classic expression of what Dan Metlay of the DOE calls the 'doughnut process' where intense local support surrounds the nuclear facilities and rapidly diminishes into a hostile surrounding region. This feature is replicated in the politics of the area. The two counties comprising the Tri-Cities (Benton and Franklin) are intensely Republican, in common with rural, agricultural eastern Washington. To the west the states of Washington and its neighbour Oregon are liberal, Democratic and environmentalist. Oregon shares the Columbia River in its lower reaches and in eastern Oregon farming and forests predominate and nurture a strong environmentalism in that state. Thus, within its state Hanford is relatively weak compared to, say, the Savannah River Site (SRS) which is situated in the less hostile political climate of South Carolina. According to Harold Heacock of TRIDEC (interview, 2013) this explains why the SRS maintained some production while Hanford lacked the political clout and support that might have enabled it to continue its nuclear activities.

Richard Morrill, writing at the turn of the century, attributed what he saw as Hanford's powerlessness to the set of power relations which privileges the regional and national scales of government over more local interests. The influence of the federal (DOE) and state governments ensured that 'the real balance of power becomes clearly regional and environmental over local and developmental' (1999, p.13). But the evidence is not conclusive and it might equally be argued that during the era of clean-up, environmental and economic interests are compatible and supported at all levels of government. Certainly, with little prospect of any further nuclear developments on the Hanford site the regional political conflicts have died away. The focus on clean-up has produced a new set of political relations, a broad consensus between hitherto opposed groups. Susan Gordon of the Alliance for Nuclear Accountability explained this as 'a change in dynamics once production ceases' (interview, 2004).

This consensus is reflected in the approach and focus of the two processes that have emerged to oversee the Hanford clean-up. The first of these is the cooperative legislative process inaugurated by the Tri-Party Agreement of 1998 between the USDOE, the federal Environmental Protection Agency (EPA) and the state of Washington's Department of Ecology (DOE). These are the bodies with the regulatory and implementation powers to ensure and enforce clean-up. The purpose of the Agreement was to establish an action plan for the treatment, storage or disposal of wastes at the Hanford Federal Facility, in particular the disposal of tank wastes, the clean-up of contamination, decommissioning of units and operation of a vitrification plant. It provides for the implementation of the relevant Acts.[2] The purpose of the Agreement is to ensure that environmental impacts are thoroughly investigated and appropriate responses are made and to provide a procedural framework with priorities, milestones and actions. In particular the Tri-Party Agreement Action Plan underlines its consensual approach, 'To facilitate cooperation, exchanges of information, and the coordinated participation of the parties in such actions' (Tri-Party Agreement, 2003, p.5). Kathy Conaway, compliance inspector with the Washington DoE, explained the Agreement is unique to Hanford and provides for an integrated management approach clearly defining responsibilities. Agreement is achieved through negotiation among the regional federal officials of the EPA and DOE and the Washington State Department of Ecology and solutions are based on compromise. The whole approach is 'professional, collaborative and constructive' (interview, 2004).

This collaborative approach is supported by a second process, the Hanford Advisory Board (HAB), which introduces an element of public participation to the Hanford clean-up. The HAB's primary mission 'is to provide informed recommendations and advice … on major policy issues related to the cleanup of the Hanford site' (Keystone Center, 1993, p.4). It has 31 members drawn from local government, local business, the Hanford work force, local environmental interests, regional environmental and public interests, public health, Native American tribes, Oregon state, universities and lay public. The Board meets in public over two days roughly every two months with the parties to the Tri-Party Agreement in attendance. It develops consensus advice through a deliberative process. The Board's advice is worked on until the point where agreement is reached so consensus means unanimity among the members. The approach is collaborative though Susan Leckband, Vice-Chair, insists 'we call their baby ugly sometimes'. Among the issues on which it has pronounced are the tank wastes, removal of spent fuel, cleaning up the wastes along the Columbia, restoration of the 300 Area, worker safety and pensions, risk based end states and so on. Given its diversity of interests the Board's advice commands respect and is regarded as influential both in general as well as on detailed issues. The HAB seems to have been remarkably successful in achieving cooperation if not cooption of anti-nuclear interests.

Throughout its history as a nuclear reservation, Hanford has remained relatively immune to the conflicts that have raged around the nuclear issue. That is not to say Hanford has been above conflict, merely that the conflicts have, on the whole, been engaged at a distance. The Hanford communities have proved resolutely pro-nuclear, proud of their original mission and determined to ensure the success of the clean-up. The end of the Cold War and opposition to the activities on the site undermined the role of Hanford and its role was reappraised. Once new goals had been agreed, hitherto conflicting interests could unite to defend the future of the area. Nevertheless, the vicissitudes experienced by the area have left their mark in a tangible vulnerability. There was a feeling in some quarters, as John Stang, formerly of the *Tri-City Herald,* expressed it that, 'Hanford is a big black hole into which you throw money and nothing comes out. We follow Congress more closely than any other city of our size' (interview, 1999). Even in its more settled and stable contemporary power relationships there remains a detectable if residual anxiety which contributes to a lingering sense of ambiguity and uncertainty about its role and identity in the modern world.

Social defensiveness – image and identity in a nuclear culture

This ambiguity can be perceived in another characteristic of peripheral communities, what might be called, in Hanford's case, a cultural defensiveness. In recent years this has been declining as the Tri-Cities have achieved greater economic diversity and a more established population with its roots in the region. Nonetheless, a cultural defensiveness can still be discerned. Liz Mattson of the Seattle based NGO Hanford Challenge described it as a sense of 'This is our land and not your concern. You don't live here. This is our community' (interview, 2013). The Tri-Cities are still on one side of the urban/rural, east/west divide in Washington state. But things have palpably changed. 'It was once inward looking, now looking outward more. It was Us and Them, but less so now'.

The Tri-Cities area has experienced a cultural sea change over the decades. 'In the old days, no one had any doubt about the rightness of nuclear technology and Hanford's mission, especially after the Second World War ended in the blinding light of Hanford's exploding plutonium' (D'Antonio, 1993, p.205). The atomic bomb was an expression of what Nye calls in the title of his book the *American Technological Sublime* (1994). It was both terrifying and irresistible, inspiring both fear and celebration. It possessed awesome power, created by humans but subjecting humans and nature to its destructive force. The sense of incomprehension, shock and amazement transcended ordinary perceptions and understandings to suggest the sublime. The bomb was created in secrecy by scientists scarcely accountable to any democratic process. The bomb unleashed devastation, death and trauma and its 'most common justifications were patriotism, the advancement of science and the protection of the free world' (Nye, 1994, p.229). It was this that instilled in the Hanford community the sense of pride in their achievement.

In the early post war years and during the Cold War period right into the 1970s, there was a close and unambiguous relationship between the reservation and the community. Wanda Munn, former Richland council member and Hanford engineer, captured the feeling. 'The community really has been a company town. We loved the company because it defended our country' (interview, 1999). The sense of identity with military defence found physical expression in various forms of aggressive symbolism. 'Atomic' is a recurrent word applied to shops and services in the area. Richland was christened 'Atomic City'. Bombing Range Road is a main thoroughfare in West Richland. Richland Senior High School proclaims itself in bold lettering over its entrance as 'Home of the Bombers', referring to its football team whose symbol is a mushroom cloud. In such ways does Hanford's

FIGURE 2.14 Hanford in symbols

Source: Author

historic achievement persist into the contemporary age although its mission has long been accomplished. The simplistic discourse of the early years with its emphasis on patriotism, defence of the nation and the necessity of the nuclear deterrent has now been supplanted by a more complex discourse reflecting the changes from secrecy to openness, from production to clean-up and from federal domination to dependence and diversification that have characterised Hanford's more recent history. Nevertheless, Hanford's history has had a formative influence on its contemporary image and identity.

Hanford's pro-nuclear discourse has been portrayed as a form of what Nadel (1995) has called 'containment culture' which was the prevailing narrative of the nuclear age. The phrase is ambiguous referring at once both to containing communism as well as to the assumptions and constraints, that is, the 'containment' of public policy discourse that were all too evident at Hanford for the first three decades after the war. To these two meanings, Kinsella (2001) adds a third, namely, the material containment of the accumulated nuclear wastes which has succeeded nuclear bombs as the focus around which the contemporary nuclear discourse of Hanford has developed.

The earlier forms of the nuclear discourse provided a grand narrative of order, secrecy, the domination of science and technology and the suppression of dissent. Sanger (1989) records the attitudes to work, place, project and people of the pioneer settlers, construction workers, designers, scientists and military personnel engaged in creating Hanford. Belief in the necessity of a project that would win the war enabled them to endure the hardships of communal life in barrack like buildings in the dusty desert of the mid-Columbia. Their passive and unquestioning confidence in the legitimacy of the nuclear project is also evident in the narratives recorded by Freer (1994). These workers evinced a strong sense of belonging and an attachment to the area and its history. These were people whose commitment was so strong that they sacrificed a day's pay to provide funds for a Boeing B17 Flying Fortress to aid the country's national defence. Like Freer, Paul Loeb in his book *Nuclear Culture* also described the early pioneers, or 'old hands', as conformist, conservative and generally single-minded. They tended to subsume the implications of their work in a 'pragmatic tractability' (1986, p.96) preferring 'to operate their reactors or raise their kids, and to leave political controversies to others. When they did express feelings, they coincided with the Area's generally pragmatic, get-the-job-done attitude' (p.97). Loeb used the term 'nuclear culture' to describe the attitudes, feelings and beliefs of the people living in the Tri-Cities region. *Nuclear Culture* (Loeb, 1986) is an anti-nuclear polemic which looks at the Hanford reservation and its communities through the eyes of workers and residents. Loeb gives a generally sympathetic portrait of the people of Hanford. He notes the strong spirit of community built up in an isolated area where everything was 'subordinate to Hanford's production rhythms' (p.89). This 'instrumental' purpose regarded anything autonomous as 'at best irrelevant and at worst an obstacle to be destroyed or altered' (p.77). By emphasising the integration between community and nuclear industry Loeb's purpose is to create an image of a place of risk, a 'demonic fountainhead, a symbol of a broader war culture' (p.101).

Of course, as both Loeb and Freer attest, the culture was not entirely homogeneous for there was some dissent, political debate and concern for social issues and there were differences, even then, between the generations. Not everyone worked in Hanford or was dependent upon it. The three cities that made up the Tri-Cities had developed their own distinctive personalities (Findlay and Hevly, 2011). But Richland especially had developed during a post-war period which was more deferential than today and was involved in nuclear weapons, an activity that demanded a disciplined, compliant workforce living in a community that was inward looking and protective of its integrity.

Some aspects of this earlier culture have survived. According to retired engineer Wanda Munn there are feelings of loss mixed with feelings of pride especially among the older residents. 'People need to understand that the life they've known is over. We are in a major transition zone' (interview, 2004). Lee Burger, one of the pioneers, recognises a loss of morale among many of the workers and also a feeling of being looked down on by outsiders. There is today a sense of profound change in the image and identity of the area which reflects its changing role in the economy and society. The sense of clarity of purpose, collective endeavour and common understanding has been eroded and largely supplanted by greater uncertainty, contested understanding and a more fragmented community. This is consistent with a shift from a modernist discourse associated with the nuclear industry to a postmodern discourse reflecting the more complex economy and diverse society that constitutes the Tri-City area today.

The shift has been associated with the move to clean-up. Kinsella (2001) argues that clean-up introduces new frames of reference in which expertise may be challenged and the nature, pace and extent of remediation may be questioned. Alongside experts, other stakeholders and members of the public (as in the Hanford Advisory Board) become involved in deliberations about the future of the area. The boundaries of deliberation are negotiated between the various interests, a process which Kinsella calls 'discursive containment'. There is still an effort on the part of the nuclear institutions to define and contain the area of debate, a position which becomes increasingly difficult to sustain as public involvement widens the boundaries of discourse.

The changing discourses surrounding Hanford have resulted in changes in its image and sense of identity, to the point where it might be argued an identity crisis had developed. In the early years Hanford possessed a confident, positive image and sense of identity as the place which made a major contribution to winning the war and securing the subsequent peacetime. In its later years as the extent of pollution emerged it acquired a negative image as the most contaminated place in the Western world. Although the stigma persists, in some ways there is the belief that the image of the area has improved as the benefits and achievement of clean-up campaigns are recognised. Hanford may still have an image problem but at least it does have an image. As Sean Stockard of TRIDEC notes, 'The only thing worse than a bad image is having no image at all' (interview, 2004). And there is a widespread feeling that the attitude of defensiveness is dissipating and disappearing in the face of a qualified optimism about the area's long-term future.

For around three decades Hanford's identity was shaped and its positive self-image supported by a pro-nuclear discourse about national security, containment and the triumph of technology. This was a powerful discourse that reinforced Hanford's own economic security and development. Gradually this discourse was undermined as alternative and oppositional ideas took root in a context of greater dissent, growing environmental concern, the opening up of Hanford's secrets and awareness of nuclear hazards. Hanford's society was changing as its economy diversified and a new generation with new attitudes and ideas began to replace the early pioneers. These changes ushered in a period of greater uncertainty, competing and conflicting images and a less assured identity. For a while an anti-nuclear discourse from without began to question the role and future of Hanford, undermining its prospects for continuing nuclear production. There followed a third phase, one of growing awareness of the problems of the site and the need for diversification. This inaugurated a new discourse of clean-up in which Hanford became a part but much less the dominant part of a more diversified, complex set of communities still with strong integrating institutions but also increasingly integrated into a wider regional context. This once isolated, self-aware and introspective area has shuffled off its frontier mentality to find a more secure and settled place in the modern world.

Environmental degradation – how clean is clean enough?

By any standards the clean-up at Hanford presents a formidable task often presented in hyperbolical terms. The wastes already on the site make it, according to former union leader Jim Watts, 'the largest waste repository in the western world' (interview, 1999). According to Shulman (1992) 'Hanford represents one of the most daunting environmental catastrophes the world has ever known' (p.94). In their scale and extent the remediation challenges facing Hanford are comparable to those experienced in the contemporary Russian Cold War reprocessing sites at Mayak (Chelyabinsk), Tomsk and Krasnoyarsk (Bradley, 1998). The scale of remedial operations at Hanford also makes it the 'the largest civil works project in world history' (Zorpette, 1996, p.88). And, it is an unending problem. From the outset the clean-up included removing high level wastes from the leaking tanks, providing secure storage for plutonium stocks, removing spent fuel from the reactor ponds on the Columbia River and decontaminating the huge reprocessing canyons. Apart from these massive projects there were the intractable problems of radioactive plumes beneath the site, contaminated soil and a myriad of problems associated with redundant facilities, waste dumps and other hazards scattered across the site. The clean-up of the high level tank wastes is especially problematic in scale and cost. 'The costs, complexity and risks of the Hanford high level waste project rival those of the US manned space programme, but have far greater potential consequences to the human environment' (Bob Alvarez quoted in Edwards, 2004). George Hinman, Emeritus Professor at Washington State University Richland, believes that leakage from the tanks is so widespread that cleaning up the aquifer presents an almost impossible task (interview, 1999).

The legacy of contamination means that a part of the site, albeit relatively small, cannot be released to unrestricted use and will require indefinite institutional control in some form. That part of the Hanford site may be considered a 'National Sacrifice Area', a place 'where environmental contamination could prove too costly and remote from human contact to merit restoration' (Shulman, 1992, p.94). Gene Schreckhise, environmental scientist at WSU, agrees that the central 200 Area at least will remain 'for ever a repository'. But, he argues, 'knowing what they didn't know in 1944 and 1945 it's amazing they didn't crap it up a whole lot more' (interview, 2004).

Despite the complexity, cost and inevitable inefficiencies, some progress with clean-up has been made (USDOE, 2004). By the early years of this century a series of projects were well under way and some of the most conspicuous problems had been tackled (Oregon Dept. of Energy, 2004). Plutonium and plutonium-bearing materials had been stabilised and packaged for secure storage on site. Most of the spent fuel from the reactor K basins on the Columbia had been removed to the central plateau and one by one the reactors were being cocooned, their graphite cores placed in interim safe storage for at least 75 years. More than 500 inactive facilities had been decommissioned, 210 contaminated sites had been cleaned up and some areas of contaminated groundwater had been treated. But, by far the biggest project was the treatment of the HLW tank wastes where progress has been slow. Liquid wastes have been removed from the single-shelled tanks in preparation for eventual vitrification. The Waste Treatment Plant (WTP) project has proved costly and troublesome while, at the same time, sustaining substantial activity at the site.

The overall plan focuses on three principal components. One is to complete the clean-up of the river corridor containing the redundant reactors along the river and the fuel fabrication, research and other facilities in the 300 area in the southern part of the site. This involves removing remaining wastes, and remediating or containing extensive areas of soil and groundwater contamination. The second is to clean up the central area of 75 square miles where wastes are concentrated and which is a complex area of waste sites, treatment and disposal facilities and 60 square miles of contaminated groundwater with plumes extending far underground. The intention is to clean up an outer area by around 2020 leaving the inner core of 10 square miles dedicated to long-term waste management and containment. The third component, clean-up of the tank wastes, is both notorious and intractable. There are problems with leakage, questions of what to do with the tank sludges and the enduring technical and cost problems surrounding the WTP. Ultimately the intention is to return cleaned up land to beneficial uses, to the Tribal Nations, for business and for recreational uses. Around half the total site area lies to the north and west of the clean-up zones and comprises the Hanford Reach National Monument devoted to conservation and preservation and including the B reactor National Historic Landmark (USDOE, 2013).

There are two fundamental but connected uncertainties about clean-up – the pace (how long will it take?) and the end state (how clean will it be?). On the first of these the timescales typically continue to lengthen. The DOE clearly has an interest in accelerated clean-up and initially believed this to be achievable 'without compromising the quality of the cleanup and in compliance with applicable requirements and cleanup standards' (USDOE, 2004, p.S.5). Although the DOE has defined its clean-up strategy it is expressed in terms of end states for different projects and parts of the site and studiously avoids offering such a hostage to fortune as precise or even expected completion dates (USDOE, 2013). Of course, extensive parts of the site will be released to end uses (recreational, ecological, industrial) reducing the site to the intractable core areas of contamination. This leaves the most costly, complex problems where the ratio of cost to risk reduction increases. There comes a point where arguably the extra cost achieves minimal improvement and may result in exposing workers to unjustifiable risk.

This relates to the second uncertainty, what is meant by clean-up? As Gephart points out , 'Agreeing a site should be cleaned and knowing what cleanup means are different' (2003, p.8.3). 'Cleanup is a negotiated, conditional state' (p.8.6). The potentially conflicting objectives over the pace and end state of clean-up have led to a wary process of detailed negotiation of one issue at a time through the Tri-Party Agreement closely scrutinised by the Hanford Advisory Board. In the absence of specific clean-up standards the DOE has been arguing for what it calls 'risk based end states'. This would involve establishing a level of acceptable risk for each site appropriate to the circumstances of the site. This flexible approach would weigh up the risks to humans and the environment against the costs and technical considerations

FIGURE 2.15 Hanford clean-up, cocooned F reactor

Source: US Department of Energy

FIGURE 2.16 Waste storage tanks being restored

Source: US Department of Energy

involved in clean-up. Put crudely, a risk-based method might range from doing nothing to scrubbing out every molecule on a site. What it certainly means is that some sites might be available for unrestricted use while others, notably the 200 Area in the central plateau, would be permanently off limits. However, a major complication for the DOE is that after nearly 25 years of clean-up, they still do not have either an acceptable or technically defensible risk-based decision making process to weigh the merits of key clean-up options (2013 interview with Gephart).

The risk-based end use concept provides an arena for controversy over some basic clean-up issues such as: should all the tank wastes be vitrified, should the reactors be shifted to the central area; should all buildings be demolished; should groundwater continue to be pumped and cleaned though it will never be clean? Critics of the clean-up process would argue that the approach is too technical, too specific, focusing on particular hazards and preventing discussion of a wider range of issues. It provides a spurious scientific precision to matters that are difficult to

FIGURE 2.17 Conservation on the Columbia Hanford Reach National Monument

Source: US Department of Energy

verify or that depend on value judgements. There will be disputes over the extent to which a situation (e.g. of contaminated groundwater) has become 'irretrievable and irreversible' (though the phrase itself is highly contested). There will be arguments over proportionality, how much should be spent to remove the last increment of risk. In practical terms these disputes involve defining how deep do you dig, how much soil do you remove and so on. Different perspectives will be brought to the problem. For the Indian tribes, Hanford is their homeland, a place of cultural and spiritual as well as material significance. They expect the area (or most of it) to be returned to unrestricted use so that they can return to collecting, hunting and fishing in their tribal lands. Other interests might be reasonably content with clean-up that allowed industrial but not residential end use in certain parts of the site (for example in the 300 area where some contamination persists). At its core the conflict is over the irresolvable question, 'How clean is clean enough?'.

The best that may be said is that clean-up has been and will be a costly, slow process with a commitment to concentrating and containing the wastes and contaminated soil and water. There is the expectation that eventually the most dangerous wastes, spent fuel and vitrified tank wastes (if the WTP is eventually commissioned) will be removed to a national deep repository should a site ever be found and a facility ever be built. But, that is something for the distant future, a generation or more hence. For the moment, Hanford's clean-up has achieved a state of relative stability, a kind of status quo as Gephart puts it. The money continues to flow in, contracts are made, projects proceed. Issues of value for money, strategy and objectives can be endlessly debated and critics point to the costliness, wastefulness, lack of strategic direction and lack of accountability. A state of inertia seems to have been reached. But, the commitment remains and the community has achieved an accommodation with Hanford where it remains 'the biggest game in town' but no longer the overweening component of the economy. Uncertainty about the longer term still exists and Hanford's claim on the national conscience may diminish in future. But, there remains a palpable sense of responsibility for the remediation and restoration of an area removed from the local population to serve a nation's deadly purposes. In the words usually attributed to Native American Chief Seattle (1780–1866): 'We do not inherit the earth from our ancestors, we borrow it from our children'.

Hanford – a continuing legacy

Hanford is one of the first nuclear industrial complexes in the world. Over the seven decades of its existence it has been exposed to profound changes dictated by external geopolitical forces. Its *raison d'être* as a nuclear arsenal which sustained its growth during the Cold War subsequently diminished to the point where its political relevance in the global context disappeared. But, the legacy of its pomp remains not only in the contaminated landscape of the mid-Columbia but in the conscience of a nation that recognises Hanford's contribution to national security. It is not just a product and victim of its role in history for Hanford has changed in other ways in response to the social, economic and political changes that have occurred throughout American society during these years. In particular Hanford's economy, society and politics are much more diversified. Nonetheless the nuclear presence persists and still provides some of the continuities that continue to shape the Tri-Cities area.

Taking the whole period of Hanford's existence, three key changes can be identified. The first and most obvious is the physical change in the area, the transformation of a pre-existing sparsely populated, rural area into a component of the United States military industrial complex focusing on the Tri-Cities. Second, has been the change in mission as the imperatives of production have been succeeded by the demand for clean-up of the contamination which that production produced. This relates to the third change which may best be described as a cultural transformation. In the early years an aggressive pride in achievement was succeeded by a growing defensiveness as its military role and popularity diminished. As its new role has become established and the area has grown and diversified so it has become more difficult to characterise its cultural characteristics. Though certainly less pervasive, the nuclear culture still plays a prominent part in the way Hanford is perceived and perceives itself.

These changes in scale, mission and culture have shaped Hanford's evolution from a peripheral community set apart from its wider regional context to a community increasingly integrated with its surrounding area. Of course, in a sense, Hanford was never peripheral in that it was at the very heart of the nation's nuclear industrial complex. On the other hand, like the other major national nuclear reservations – Savannah River, Oak Ridge, Los Alamos – it was remote, defended, inward looking and secretive. Over the years Hanford's peripherality has diminished as changes in each of the five peripheral characteristics have brought it more into the fold. It is now geographically less remote as it has become integrated into regional and national communications systems. Economically, Hanford is no longer monocultural but increasingly diversified. And, though it remains politically staunchly Republican and thus distant from the western part of Washington state, the political divisions that once existed between east and west, urban and rural have diminished. Culturally, too, the area has shed its combination of combativeness and defensiveness and manifests a greater confidence and security. Finally, although the Hanford site is notoriously contaminated and degraded it is being progressively improved with the prospect of all but the core returning to non-nuclear uses. Meanwhile, the Tri-Cities area has developed a reputation as a recreational and retirement area. Of course, some peripheral characteristics remain, notably a continuing reliance on federal appropriations. But, the isolation, anxiety and political vulnerability that were briefly experienced when the production mission collapsed have gone, as a pride and purpose in clean-up has developed.

The speed, scale and secrecy with which Hanford was conceived and developed were, quite simply, awesome. Its ability to survive, to withstand the vicissitudes of its subsequent history, is also compelling. Hanford's resilience, its economic survival, its social stability and its political influence relate to a geographical and cultural identity that whilst changing is distinctive and continuing. According to Michele Gerber it is a special place, as significant as Gettysburg; it is one of the places that changed the course of world history.

> Yet the immediacy of history, the demand to participate and be heard, the fears and hopes expressed, and the constant press of those coming to hear the Hanford story bring a cacophony of voices and visitors to this remote place.
>
> *(Gerber, 2001, p.2)*

In the process of clean-up the complexities and changes of society are revealed – the clinging to the old, the desire for the new. As clean-up proceeds and the generations move on so the transformation of the site will be complete. But, the part that Hanford played in history will not be so easily forgotten and will continue to affect the area in the years to come.

Visits and interviews

The following people were interviewed during visits to Hanford and Washington D.C. and the author is very grateful for their help in providing much of the substance for this chapter.

October 1999 visit to Hanford

Interviewed: Ben Bennett, Chief Executive, Port of Benton; Brian Costner, consultant, Seattle; Jerry Gough, historian, Washington State University at Pullman; Larry Haler, Mayor of Richland; Geoff Harvey, Public Relations, BNFL, Hanford; George Hinman, Emeritus Professor, Washington State University at Tri-Cities; Ron Kathren, Washington State University at Tri-Cities; Mike Lawrence, Director, BNFL, Hanford and former Hanford site manager; Wanda Munn, retired engineer and Richland council member; Lloyd Piper, former Hanford site deputy manager; Dean Schau, regional economist; Gene Schreckhise, Washington State University at Tri-Cities; John Stang, journalist, *Tri-City Herald*; Sam Volpentest, Tri-City Industrial Development Council; Jim Watts, labour leader.

March 2004 visit to Hanford

Interviewed: Carl Adrian, Tri-City Industrial Development Council; Pamela Brown, City of Richland; Greg de Bruler, Columbia Riverkeeper, NGO, Hanford Advisory Board; Lee Burger, retired engineer; Colleen Clark, USDOE, Richland; Shelley Cimon, Hanford Advisory Board; Kathy Conaway, Washington State Department of Ecology; Suzanne Dahl, Washington State Department of Ecology; Ron Gallagher, President, Fluor Hanford; Michele Gerber, Communications, Fluor Hanford; Harold Heacock, Tri-City Industrial Development Council; Susan Hughes, Hanford Advisory Board; Todd Martin, Chair, Hanford Advisory Board; Eric Olds, USDOE, Office of River Protection; Gerry Pollett, Heart of America Northwest, Hanford Advisory Board; Richard Raymond, CH2M HILL; Gene Schreckhise, Washington State University at Tri-Cities; John Stanfill, Nez Perce Tribe, Hanford Advisory Board; Sean Stockard, Tri-City Industrial Development Council; Leon Swenson, Consultant, Hanford Advisory Board; Nancy Uziemblo, Washington State Department of Ecology; Don Woodrich, USDOE, Office of River Protection.

May 2004 visit to Washington D.C.

Interviewed: Bill Barnard, CEO, Nuclear Waste Technical Review Board; Lake Barrett, Consultant; Michele Boyd, Public Citizen NGO; Margaret Chu, Director, Office of Civilian Radioactive Waste Management, USDOE; Paul Dickman, Senior Technical Policy adviser, USDOE; Susan Gordon, Alliance for Nuclear Accountability (Seattle); Brendan Hoffman, Public Citizen NGO; Michael Mariotte, Nuclear Information and Resource Service; Dan Metlay, Senior Professional Staff, Nuclear Waste Technical Review Board.

May 2013 visit to Seattle and Hanford

Interviewed in Seattle, May, 2013: Tom Carpenter, Hanford Challenge; Meredith Crafton, Hanford Challenge; Liz Mattson, Hanford Challenge; Katherine Sane, Hanford Challenge.

Interviewed in Hanford: Pamela Brown, Hanford Communities; Bill Dixon, engineer; Roy Gephart, environmental scientist, author of *Hanford, A Conversation*; Michele Gerber, historian, author of *On the Home Front*; Susan Leckband, Vice-Chair, Hanford Advisory Board; Gary Petersen, Head of TRIDEC.

May 2013 visit to Washington D.C.

Interviewed: Diane D'Arrigo, Nuclear Information and Resource Service; David Blee, Nuclear Infrastructure Council; Linda Penz Gunter, Beyond Nuclear; Arjun Makhijani, Institute for Energy and Environmental Research; Nigel Mote and Karyn Severson, Nuclear Waste Technical Review Board; Jeff Williams, US Department of Energy.

At various times I have met with officials from the Nuclear Regulatory Commission, the US Department of Energy, the Environmental Protection Agency and the Nuclear Waste Technical Review Board. I am grateful to these and others not recorded here for the range of insights they provided.

Notes

1   Gephart (2003) records that between 1944 and 1970 about 10M curies of radioactivity were discharged to the air from the first eight reactors with a further 2M later released from N reactor. Between 1944 and 1972 about 20M curies were released by the reprocessing plants (p.5.50).
2   Federal Resource Conservation and Recovery Act, Comprehensive Environmental Response, Compensation and Recovery Act (CERCLA), otherwise known as 'Superfund' and the Washington State Hazardous Waste Management Act.

References

Abbott, C. (1998) 'Building the Atomic Cities: Richland, Los Alamos and the American Planning Language', in Hevly, B. and Findlay, J. *The Atomic West,* Seattle and London, University of Washington Press, pp.90–115.

Blowers, A., Lowry, D. and Solomon, B. (1991) *The International Politics of Nuclear Waste,* London, Macmillan.

Bradley, D.J. (1998) *Behind the Nuclear Curtain: Radioactive Waste Management in the Former Soviet Union,* Columbus, Ohio, Battelle Press.

Columbia River Exhibition (undated) *ABC Homes: the Houses that Hanford Built,* 95 Lee Blvd., Richland, Washington.

D'Antonio, M. (1993) *Atomic Harvest: Hanford and the Lethal Toll of America's Nuclear Arsenal,* New York, Crown Publishers, Inc.

Del Tredici, R. (1987) *At Work in the Fields of the Bomb,* New York, Harper and Row.

DeVoto, B. (1953) *The Journals of Lewis and Clark,* Boston, Houghton Mifflin Company.

Edwards, R. (2004) 'US nuclear clean-up carries major risks', *New Scientist,* 25 July.

Ficken, R. (1998) 'Grand Coulee and Hanford: the atomic bomb and the development of the Columbia River', in Hevly, B and Findlay, J., *The Atomic West,* Seattle and London, University of Washington Press, pp.21–38.

Findlay, J.M. and Hevly, B. (2011) *Atomic Frontier Days; Hanford and the American West,* Seattle and London, University of Washington Press.

Fowler, R. and Scott, M. (2009) *Hanford and the Tri-Cities Economy: Historical Trends 1970– 2008,* prepared for US Department of Energy, Pacific Northwest National Laboratory, Richland, WA 99352, October.

Freer, B. (1994) 'Atomic pioneers and environmental legacy at the Hanford site', *Canadian Review of Sociology and Anthropology,* 31, 3, pp.305–324.

Gephart, R. (2003) *Hanford, A Conversation about Nuclear Waste and Cleanup,* Columbus, Ohio, Battelle Press.

Gerber, M. (1992) *On the Home Front: The Cold War Legacy of the Hanford Nuclear Site,* Lincoln and London, University of Nebraska Press.

Gerber, M. (1993) *The Hanford Site: An Anthology of Early Histories,* Richland, Washington, Westinghouse Hanford Company.

Gerber, M. (2001) 'Epilogue' for *On the Home Front: The Cold War Legacy of the Hanford Nuclear Site,* for the Second Edition, Lincoln and London, University of Nebraska Press.

Gerber, M. (2007) 'Epilogue' for *On the Home Front: the Cold War Legacy of the Hanford Nuclear Site,* for the Third Edition, Lincoln and London, University of Nebraska Press.

Goldberg, S. (1998) 'General Groves and the Atomic West: the making and the meaning of Hanford', in Hevly, B. and Findlay, J. *The Atomic West,* Seattle and London, University of Washington Press, pp.39–89.

Gusterson, H. (1998) *Nuclear Rites: A Weapons Laboratory at the End of the Cold War,* Berkeley, University of California Press.

Greenberg, M, Miller, K.T., Frisch, M and Lewis, D. (2003) 'Facing an uncertain economic future: environmental management spending and rural regions surrounding the U.S. DOE's nuclear weapons facilities', *Defence and Peace Economics,* 14 (1), 85–97.

Hanford Communities (2013) *Hanford from the Highway Audio Tour,* Hanford Communities, Department of Ecology, State of Washington, Lockheed Martin. Also Video Tour available.

Hein, T. (2003) *Atomic Farmgirl,* New York, Mariner Books.

Hevly, B. and Findlay, J. (1998) *The Atomic West,* Seattle and London, University of Washington Press.

Keystone Center (1993) *Convening Report on the Establishment of an Advisory Board to Address Hanford Cleanup Issues,* Keystone, Colorado, October.

Kinsella, W. (2001) 'Nuclear boundaries: material and discursive containment at the Hanford nuclear reservation', *Science as Culture,* 10, 2, 163–194.

Kuletz, V. (1998) *The Tainted Desert: Environmental and Social Ruin in the American West,* New York, Routledge.

Lee, K. (1989) 'Columbia River Basin – experimenting with sustainability', *Environment,* 31 (6), 7–11, 30–33.

Loeb, P. (1986) *Nuclear Culture: Living and Working in the World's Largest Atomic Complex,* Philadelphia, New Society Publishers.

Morrill, R. (1999) 'Inequalities of power, costs and benefits across geographic scales: the future uses of the Hanford reservation', *Political Geography,* 18, 1–23.

Nadel, A. (1995) *Containment Culture: American Narratives, Postmodernism, and the Atomic Age,* Durham N.C., Duke University Press.

Nye, D. (1994) *American Technological Sublime,* Cambridge, Massachusetts Institute of Technology Press.

Oregon Department of Energy (2004) *Hanford Cleanup: The First 15 Years,* October, Oregon Department of Energy, Salem, OR.

Perry, C.A. (1929) 'The neighbourhood unit, a scheme of arrangement for the family-life community', in *The Regional Survey of New York and its Environs,* Vol. 7, New York, pp.22–140.

Pope, D. (1998) 'Antinuclear activism in the Pacific Northwest: WPPSS and its enemies', in Hevly, B. and Findlay, J. (eds.) *The Atomic West,* Seattle and London, University of Washington Press, pp.236–254.

Raban, J. (1997) *Bad Land: An American Romance,* London, Picador.

Sanger, S. (1989) *Hanford and the Bomb: An Oral History of World War II,* Seattle, Living History Press.

Shulman, S. (1992) *The Threat at Home: Confronting the Toxic Legacy of the US Military,* Boston, Beacon Press.

Tri-Party Agreement (2003) *Hanford Federal Facility Agreement and Consent Order,* United States Department of Energy, Washington State Department of Ecology, United States Environmental Protection Agency, 89-10-REV.6.

UK Nirex (2004) Briefing Note for visit to Hanford, unpublished.

USDOE (1995) *Closing the Circle on the Splitting of the Atom: The Environmental Legacy of Nuclear Weapons Production in the United States and What the Department of Energy is Doing About It,* Washington, D.C., USDOE Office of Environmental Management.

USDOE (2004) *Final Hanford Site Solid (Radioactive and Hazardous) Waste Program Environmental Impact Statement Richland, Washington,* Richland Operations Office, January.

USDOE (2013) *Hanford Site: Cleanup Completion Framework.* Richland Operations Office, Richland, Washington, January.

Zorpette, G. (1996) 'Hanford's nuclear wasteland', *Scientific American,* May, pp.88–97.

3

SELLAFIELD, UK

A paradox of power

A dangerous necessity

From a distance, Sellafield, the largest industrial site in the UK, seems both incongruous and mysterious. The last time I visited, in 2014, I caught my first glimpse of the plant from high up in the Lakeland Fells in the declining light of a lowering, late autumn evening. My vantage point from a winding mountain road was windswept; above, the scudding clouds and below the land stretching westwards to the distant Irish Sea. To the north was the rugged bastion of the central mountain massif of the Lake District, dominated by Scafell Pike, England's highest peak, half hidden in the mist. All around was the lonely beauty of hills with dry stone walls, bracken, tumbling rills and hardy sheep. Far below, at the foot of Eskdale, I could discern the nuclear plant, occasionally lit up by flashes of sunlight through parted rain clouds. From that distance, in that eerie twilight and primitive landscape, Sellafield nestled innocently on the plain in between mountains and sea, for all the world recalling to mind the silhouette and environs of a medieval monastery.

Even up close, the purpose of this jumble of concrete and steel buildings, masses of pipework, 28 miles of road and eight miles of railway with the river Calder running through the middle, is difficult to fathom. It seems extraordinary to think that within a mere 1.5 square miles (2 sq. km) are concentrated the full gamut of the UK's nuclear activities from its post-war beginning to the present day. Here, around two-thirds by radioactivity of all the radioactive wastes in the UK are stored. They comprise all the high level waste (HLW) and most of the spent fuel which together, though only comprising 2% of the volume, account for 92% of the radioactivity of wastes stored at Sellafield (CoRWM, 2006). Sellafield hosts almost half the volume of intermediate level waste (ILW) while the bulk of the nation's low level wastes (LLW) are disposed of in the nearby low level waste repository at

FIGURE 3.1 Sellafield

Source: Sellafield Ltd

Drigg. Sellafield has a stockpile of 123 tonnes of plutonium, rising to an estimated 140 tonnes (24 tonnes foreign owned) by around 2020, by far the world's largest, which contributes around 5% of the total radioactivity. There are also stores of uranium in various forms, high in volume but relatively low in radioactivity.[1]

The iconic architectural profile of Sellafield is gradually changing as some of its earlier legacy buildings are dismantled. Already gone are the four cooling towers of Calder Hall, built initially to supply spent fuel for military reprocessing but also the first nuclear power station to provide electricity for the national grid from 1956–2003. The Windscale Piles which produced the plutonium for the UK's atomic bomb have been decommissioned and one of its chimneys demolished. The other is being gradually and carefully taken down including the bulky filter at the top, foolishly known as 'Cockroft's Folly' after the scientist who insisted it was built. His foresight proved invaluable when the chimney trapped the bulk of the radioactivity released during the Windscale meltdown in 1957. Without it the devastation might have been incalculable. Nearby is the 'golf ball' shape of the Windscale Advanced Gas Cooled Reactor (WAGR), prototype of the country's fleet of fourteen AGRs at seven sites, now decommissioned and soon to be demolished.

Scattered around the site are the buildings, randomly located cheek by jowl, containing the myriad processes, services and activities that constitute the Sellafield complex. Reprocessing, the chemical separation of plutonium and uranium from spent nuclear fuel, which was the original *raison d'être* of Sellafield, still continues. The Magnox reprocessing plant with its landmark tall chimney has operated since

FIGURE 3.2 Calder Hall nuclear power station. Inset: Queen Elizabeth at the opening, October 1956

Sources: Sellafield Ltd (main picture), Whitehaven News Archives (inset)

1964 and will cease around 2020 when the last consignments of Magnox spent fuel from the nine first generation Magnox power stations and the two military/civil stations at Calder Hall and Chapelcross over the Scottish border have been processed. Fuel from the country's seven Advanced Gas Cooled Reactors (AGR) power stations and the sole Pressurised Water Reactor (PWR) station is processed in the Thermal Oxide Reprocessing Plant (THORP), a vast and vastly expensive tall and long building, the single largest construction project of its kind in the UK and once the pride of Sellafield which began operations in 1994 with expectations of foreign orders to fill its 800 tonnes capacity. It has operated well below throughput and is now eking out its last years before closure in 2018.

The main focus of activity at Sellafield has shifted from production to clean-up and especially to clearing up the legacy from the early days of the nuclear programme when little attention was paid to dealing with the wastes. Hemmed in within the site are large grey anonymous buildings containing often unrecorded mixtures of fuel, skips and other highly radioactive debris tipped into ridges and valleys of sludge and water. Here are B29 (B refers to Building) an open single skinned storage pond, B30 'Dirty Thirty', by some considered 'the most dangerous industrial building in Europe' (Roche, 2015), the first generation Magnox pond with a combination of wastes, corroded debris and fuel in a congested area of the site and B38, the pile fuel storage pond, also in contention for the soubriquet of the most hazardous site in Europe where early spent fuels were stored uncharacterised. Together with other buildings containing swarfs and fuel cladding among other wastes, these long neglected ponds and silos have been described by the National Audit Office as posing an 'intolerable risk' and the formidable and expensive challenge of dealing with them is Sellafield's top priority for risk and hazard reduction (NAO, 2012, p.2).

The sheer complexity and scale of operations crammed on to such a small site is incomprehensible. Sellafield is tightly defended by double rows of fencing, heavy barriers at the entrances and armed police patrols. Shut off from the outside world it has nurtured a reputation for secrecy that is increasingly challenged in an

FIGURE 3.3 Windscale Advanced Gas Cooled Reactor

Source: Sellafield Ltd

FIGURE 3.4 Legacy waste pond at Sellafield

era that professes openness, transparency and accountability. The attention paid to safety and security is comprehensive and meticulous but the radioactive discharges, accidents and major incidents that have occurred over the years have made Sellafield a notorious and controversial plant. That it is hazardous is an inescapable fact. The probability of a major incident may be small but the possibility persists. I felt it myself on a visit back in 2004 when standing on a platform above the massive concrete shield below which were highly active liquor (HAL) tanks containing 99% of the radioactivity from spent nuclear fuel. Maintaining the controlled storage of HAL through cooling, agitation and monitoring before it is vitrified 'presents a unique waste management challenge' (Sellafield Plan, HAL Workstream, 2011, p.3). I turned to my colleague, a renowned radiation scientist, and asked him how safe we were. He looked up at the miles of cables and pipes above us indicating their exposed vulnerability in the event of disruption which could affect the cooling of the liquors below releasing a massive burst of radioactivity. While such a scenario did not seem possible, the thought was chastening. 'You could say we are standing on the most dangerous place on earth' he remarked. The hyperbole encapsulated the idea of Sellafield as a dangerous necessity. It is also a place where, of necessity, people live and work and it is the relationship between Sellafield and the community of West Cumbria that is the story of this chapter.

The chapter falls naturally into three parts. The first, following a historical narrative, describes the post-war origins of Sellafield as the centre for plutonium production for the UK's nuclear weapons programme and subsequently its role at the heart of the country's civil nuclear complex. After its early unchallenged expansion, Sellafield's role in reprocessing was increasingly challenged and it became a focus for conflict most notably over the THORP project. The second part traces the transition from Sellafield's original purpose of production to an emphasis on clean-up and waste management. The role of Sellafield as both the source and potential solution for the long-term management of the UK's nuclear legacy is portrayed through the attempts to find a site for the disposal of the most dangerous wastes. Throughout the first two parts the theme of changing discourses provides an underlying reflection and explanation of the changing power relations that have shaped the social as well as the physical components of Sellafield's nuclear legacy. In the third part these components are examined in an attempt to explore Sellafield in its geographical, social, economic and political context as a place on the periphery but, in terms of managing the nuclear legacy, a place at once both powerless and powerful.

The research for this chapter benefited from the access I have had to materials and decision makers through my various political, governmental and non-governmental roles over the years. But I have also drawn on the conversations I have had during many visits to Sellafield and the people I have met are recorded at the end of the chapter.

From Trust in Technology to Danger and Distrust

Sellafield – the nuclear heartland

Sellafield is the heart of the UK's nuclear industry. Its first nuclear plant, the Windscale Piles, produced the weapons grade plutonium which could be recovered from spent fuel in associated chemical separation (reprocessing) works. In its first years Sellafield[2] (or Windscale as the site came to be called from 1947 to 1981 before reverting to its original name) was dedicated to the atomic weapons programme leading to the detonation of the first British nuclear bomb off the northwest coast of Australia in October 1952 (Milliken, 1986). The mystery and secrecy that has been a feature of Sellafield's nuclear culture stems from its military origins and still, in some respects, persists to this day.

Sellafield also became the fountainhead of the nascent civil nuclear programme with the development of two dual purpose (military and civil) nuclear power stations, one on the site at Calder Hall (Jay, 1956), the other at Chapelcross across the Scottish border, both close to the reprocessing facility at Windscale. The emphasis was firmly on the peaceful purpose of producing electricity manifested in the Queen's speech on opening Calder Hall in October 1956. There is, in retrospect, an innocent irony in the millennial tone of the speech. 'Future generations will judge us, above all else, by the way in which we use these limitless opportunities, which Providence has given us and to which we have unlocked the door'. In those heady days Sellafield was the apotheosis of the Trust in Technology discourse which held scientific achievements in awe. 'It was a golden age of public acceptance' (Herring, 2005, p.68). A sense of heroic participation in a vital project is evident in the reflections of workers like Tom Tuohy interviewed by Jean

McSorley in *Living in the Shadow* (1990): 'Things that were done down there in the early days were absolutely wonderful, I enjoyed my working life, it was tremendous. It was really exciting' (p.200). Wynne et al. in their study of Sellafield summarise the spirit of the period,

> The post-war period in which the establishment of the UK nuclear industry took place facilitated the easy adoption of views incorporating: a modernistic vision of an oncoming nuclear age; an unambiguous confidence in the ability of science to solve technical problems; a sense of mission, pride and achievement in the tasks involved; and lastly, a paternalistic spirit extending to the patronage of the local community.
>
> *(Wynne et al., 1993, p.17)*

This pride took a severe jolt in 1957 when the Windscale Piles caught fire. At the time, the fire was put out by the heroic but makeshift efforts of workers taking huge risks as they poked the flaming fuel elements out of the channels to safety. Although the 'incident' (the word commonly used by the nuclear industry to refer to an accident) was made public, the full extent of the damage and danger was not fully explained as the *New Scientist* indicated at the time,

> Public confidence has been severely shaken by what appeared to be attempts to minimise the gravity of what had taken place at Windscale, and even more by the extremely late hour at which any precautions to safeguard public health were put into effect. The escape of fission products may have been small, but night calls by police two days after the first leaks occurred suggest an unfortunately belated awakening to the degree of contamination that might in fact be involved.
>
> *('Windscale Fire', 17 October 1957, reprinted 18 October 2006)*

Indeed, there were no evacuations or other precautions taken and little information was provided. Although the state of emergency was obvious in the local area where milk was poured away and people were forbidden to drink it, its impact on the wider public was limited since 'no-one was told not to drink milk in the South' (Jean McSorley, interview, 2014). Yet, Windscale was a major emergency and the escape of radioactivity through the chimney before the fire was brought under control led to widespread contaminaton of the environment and to milk supplies (Breach, 1978). Indeed, had it not been for the filter on the chimney ('Cockcroft's Folly') which trapped 90% of the radioactivity, the accident might have been truly catastrophic. Like the even lesser known but more devastating accident at Kyshtym near the Mayak reprocessing plant in the Urals also in late 1957 (Medvedev, 1979; Brown, 2013), the scale and the possible consequences of the Windscale fire were not revealed until much later. These early accidents made little impact on the public consciousness at the time and did not disturb the settled discourse. But, the opening of Calder Hall and the Windscale fire established Sellafield's contradictory role in the nuclear industry as 'both the technological keystone of that industry and its Achilles' heel' (Bolter, 1996, p.2).

Unwitting deference to expertise encouraged a culture of secrecy which experts regarded as an entitlement and which citizens saw no need to challenge. During the decades that followed Windscale, the UK's first generation nuclear energy programme got under way with the publication of the White Paper *A Programme of Nuclear Power* in 1955 (Ministry of Fuel and Power). Nine power stations, eight on the coast and one inland (Trawsfynydd, Snowdonia), were developed using the Magnox technology pioneered at Calder Hall. By the 1970s the second generation of seven Advanced Gas Cooled (AGR) stations based on the prototype Windscale AGR (affectionately known as 'Wagger') had been commissioned, though the programme took many years to complete. Three AGRs were sited alongside Magnox stations and four were on greenfield sites including two close to built up areas partly to persuade the public 'to accept nuclear reactors as just another form of nuclear activity' (Openshaw, 1986, p.136). Yet, while the Magnox stations had been built apparently without a murmur of dissent, opposition was developing and notably on a large scale around the Torness AGR site in East Lothian, Scotland and, to a lesser extent, around the putative sites (not chosen) in Cornwall (Herring, 2005; Walsh, 2000). That is not to say that opposition had been hitherto dormant, quite the contrary as Horace Herring strenuously points out. But, it is fair to say that it was during the 1970s that nuclear power became the focus of growing public concern about the risks, economic and environmental, associated with such a large-scale commitment to the technology. From a series of evanescent campaigns was born a more focused opposition which had all the makings of an anti-nuclear movement albeit more subdued than the mass protests in contemporary France and Germany (see Chapters 4 and 5). Its moment came at Sellafield with the proposal for a new reprocessing works there.

Sellafield – the nuclear laundry

Reprocessing was, of course, already well established at Sellafield, indeed was its *raison d'être*, an integral part of the nuclear fuel cycle, processing spent fuel from the Magnox programme. It was the key element in the genesis of the military and civil nuclear complex secretly and uncontroversially built at Sellafield in the early post-war years. But, the political and cultural environment of the 1970s was altogether different. It was a more questioning, certainly more sceptical if not cynical time when the establishment, scientific and political, was being challenged and, to a degree, held to account. In the nuclear sector doubts were accumulating as the potential dangers became more evident and as nuclear energy faced challenges of cost, delays and competition. In short, the shift to a new discourse of Danger and Distrust was under way although it did not become full-blown in the UK until the 1980s in the aftermath of the Three Mile Island disaster in the United States in 1979. But, the main components of the discourse were fully evident during the inquiry into the Thermal Oxide Reprocessing Plant (THORP) designed to

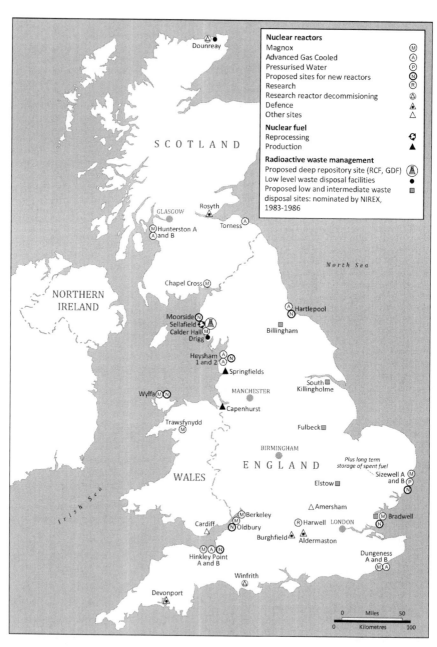

FIGURE 3.5 Map of nuclear facilities in the UK

Map by: John Hunt

produce plutonium for, among other uses, fast breeder reactors (under experimental development at Dounreay in northern Scotland) and to make money for the UK by reprocessing fuels from overseas. This is not the place to discuss in detail the purpose, viability or performance of THORP, save to say that, on all three counts, it has not met its original envisaged intentions. Its role in the Sellafield story is significant for it marked the point of discursive transition towards conflict over nuclear energy. Specifically the conflict focused around the THORP public planning inquiry. Lasting 100 days during 1977 it was the longest running planning inquiry up to that point, to be exceeded later only by the Sizewell B PWR inquiry (365 days) though both the nuclear inquiries were subsequently overhauled by the London Heathrow Terminal 5 inquiry at 524 days. The Windscale inquiry was the first significant contest between pro- and anti-nuclear factions using the quintessentially legalistic and adversarial British method of formal examinaton and interpretation of the evidence to determine the outcome.

Although Magnox reprocessing in a dedicated plant (B205) had long been a routine function of Sellafield, what was being proposed was a new dimension in terms of scale and purpose. The new plant was envisaged to have a capacity of over 1,000 tonnes per year though the intended throughput would be considerably less at around 600 tonnes and the actual throughput rather less than that (Walker, 2007). The idea was to dedicate half the throughput for the first ten years to processing the UK's AGR fuel and half to overseas customers with the option of repatriating the wastes arising. At the time, the UK's nuclear programme was expected to undergo considerable expansion with the completion of the AGR fleet

FIGURE 3.6 The Thermal Oxide Reprocessing Plant (THORP) at Sellafield

Source: Sellafield Ltd

and later the introduction of a generation of PWRs, beginning with the first at Sizewell B (which would also prove to be the last nuclear station in the UK for at least a generation and, in any event, its spent fuel would be stored on site rather than reprocessed). Reprocessing would help to augment the stockpile of plutonium and uranium which could be converted into mixed oxide fuel (MOX) for both fast breeder and conventional thermal reactors. At the same time there appeared to be considerable profit to be made from undertaking reprocessing for foreign customers, notably Japan, Germany and Switzerland.

Initially, the project had government support but opposition began to develop especially as incidents of leakage and accidents from existing facilities at Sellafield were revealed. Sellafield's image was further damaged by the *Daily Mirror* which on its front page portrayed the importing of foreign spent fuel as a 'Plan to Make Britain World's First Nuclear Dustbin' (21 October 1975). The economic benefits were also being questioned as a range of opponents, including national NGOs like Friends of the Earth, Greenpeace and the Town and Country Planning Association together with local groups including Cumbrians Opposed to a Radioactive Environment (CORE), presented evidence to the inquiry. The grounds for opposition included: dangers of international trade in nuclear materials; the development of an energy economy based on fast breeder reactors; issues of security in a plutonium economy; the economics of the project; and the environmental hazards it presented. The arguments, assumptions and assertions put forward on behalf of the developer, British Nuclear Fuels Limited (BNFL), ultimately won the day and the Planning Inspector, Justice Parker, provided a favourable judgement and pronounced that permission should be granted 'without delay' (Breach, 1978, p.141) and the plant was given the go-ahead in 1978.

The inquiry had pitched experts in combat, the arguments had been conducted at an 'elite level' (Susan Meyer, interview 1994). It was more related to the international and national context than a local one though it did have an impact on the burgeoning environmental movement. In the following years the nuclear issue, and especially the role of Sellafield, became intensely politicised. In particular, 1983 can be identified as the pivotal point when nuclear issues achieved widespread public awareness and mobilisation of anti-nuclear groups. It was, in Jean McSorley's words, a 'huge' year. 'In 1983 it just went ballistic through all the various things' (interview, 2014). This was the year during which, in the summer, international protests caused the annual sea dumping of nuclear waste into the Atlantic Ocean to be abandoned. It was the year when, in early autumn, Nirex (the Nuclear Industry Radioactive Waste Executive) announced its intention to develop radioactive waste disposal facilities at Billingham on Teesside and Elstow in Bedfordshire (and later at three other sites in Eastern England) thereby fomenting opposition in areas hitherto untouched by nuclear's shadow (Blowers et al., 1991). It was also the year that put Sellafield into the spotlight. Yorkshire TV transmitted to the nation its documentary, 'Windscale – the nuclear laundry', on 1 November drawing attention to the possible link between nuclear installations and leukaemia clusters. Shortly afterwards, divers acting for *Greenpeace* attempted

FIGURE 3.7 Monitoring the Seascale beach after radioactive discharge in November 1983

Source: therealgaffney.zenfolio.com

to block the pipeline from which radioactive 'crud' had been deliberately discharged into the Irish Sea, some of which had washed ashore leading to the closure of the beach for 24 hours and warnings not to use beaches extending 20 miles in each direction from the Sellafield plant.

Certainly the impact of 1983 was, for Sellafield, transformative.

> Sellafield, at one time at the forefront of nuclear technology anywhere in the world – a symbol testifying to intense positive identification with nuclear technology – was for many people, by 1983, a byword for the dirty end of a dangerous industry.
>
> *(Macgill, 1987, p.12)*

One of the consequences of that fateful year was the investigation into possible increases in cancer in West Cumbria conducted by Sir Douglas Black which duly reported, inevitably and ambiguously, that the link was 'not one which can be categorically dismissed, nor on the other hand, is it easy to prove' (Black Report, 1984). While this seemed to give a qualified reassurance, the uncertainties surrounding risk at Sellafield contributed to a deepening antipathy to nuclear power during the 1980s, a widening gulf between the nuclear industry and the public leading to what might be called a 'crisis of legitimation' (Macgill and Phipps, 1987) and a concomitant decline in public support for nuclear power.

THORP almost fell vicitim to nuclear's diminishing fortunes. During the 1980s the plant, the largest construction project in Europe at the time, was gradually emerging employing thousands of construction workers and confirming Sellafield's growing dominance in the West Cumbrian economy. During the period of its construction, economic circumstances had become much less favourable. The UK nuclear programme was on the wane and the environmental and security issues aroused by growing public anxiety, especially heightened in the aftermath of Chernobyl, had removed any remaining lustre from nuclear development. More specifically for THORP, not all domestic AGR fuel was committed, while the expected levels of foreign business had not materialised although the plant had become more dependent on them for its projected throughput and profits. The Breeder programme had collapsed and costs had escalated. A related issue was the fate of the Sellafield MOX plant (SMP) built to make fuel for both UK and foreign customers which opened in 2001 with a capacity of 120 tonnes per year. It only produced 5 tonnes in its first five years and was declared failed and shut down in 2011.

Sellafield – an international issue

The argument about THORP was no longer about the size of the benefits it would bring; rather it was about whether the plant would become an economic and environmental liability (Breach, 1978; Hall, 1986; Aubrey, 1993; Friends of the Earth, 1999; CORE, undated). In these circumstances, and with the plant nearing completion, an attempt was made to prevent its opening. It still had to be authorised by the regulators and this opened up an opportunity for further challenge and judicial review, a tactic initiated by Greepeace and used in later conflicts. An independent report commissioned by the government eventually concluded that the economic benefits of operation and the loss of revenue from overseas if it were to be abandoned would be significant. The subsequent chequered history of THORP in terms of incidents, lost contracts, performance and purpose suggests strongly that it might well have been better had it never opened at all. In 2005 a leakage of radioactive waste from a cracked pipe into a sump was discovered leading to closure of the plant until January 2007. More and more THORP's function was shifting from that of manufacture (of plutonium) to that of waste management as it took in spent fuel from AGRs and foreign customers and produced waste that, for the time being, had nowhere else to go. So, by default THORP's real function has become that of Sellafield as a whole: clean-up and waste management. This transformation from production to clean-up will be complete in 2018 when reprocessing at THORP finally closes.

Another potentially controversial aspect of THORP was its role in reprocessing and managing spent fuel and radioactive wastes arising from overseas customers

(Blowers and Lowry, 1993). Both the plutonium and the resulting wastes were to be repatriated with the variation that a radiological equivalent additional amount of high level waste (HLW) could be returned in lieu of the large volumes of intermediate and low level wastes (ILW, LLW) derived from reprocessing (known as 'substitution'). Thus a smaller volume of substituted vitrified waste would be repatriated thereby requiring far fewer shipments back to the country of origin. This arrangement implies that some foreign wastes (ILW and LLW) will remain in the UK, mainly at Sellafield. Similarly, some of the foreign-owned spent fuel already imported may not be reprocessed before THORP closes in 2018 and so will, according to the government, be managed in the UK 'by means of interim storage pending disposal, taking ownership where necessary' (DECC, 2015a). As for returning foreign wastes, although the UK remains committed to repatriation, very little had been repatriated by 2015 and it was planned to export a total of over 1,500 canisters of HLW (263 cubic metres) over ten years (DECC, 2015a). It may be doubted that this programme will ever be completed especially given the risks from transporting the wastes and the difficulties of finding ways of long-term management in the countries of origin, notably in Germany (see Chapter 5).

There is also the issue of what is to be done with the foreign-owned plutonium amounting in 2013 to around 24 tonnes out of a total UK stockpile, the world's largest, estimated to reach 140 tonnes at the conclusion of reprocessing. The preferred strategic policy for the management of plutonium stocks in the UK was identified in 2011 as reuse in mixed oxide fuel (MOX) (DECC, 2011). Further exploration of options by the Nuclear Decommissioning Authority (NDA) included the reuse in advanced or fast reactor systems. However, even assuming markets could be found (a quite heroic assumption), reuse would require the development of a new MOX facility at Sellafield and the development of new technologies to a point where they are credible options, both a somewhat distant, and it might be added unrealistic, prospect. Meanwhile, the options of long-term storage (and eventual disposal) or the variant of immobilisation and direct disposal need also to be pursued. Even if the reuse options materialise, foreign plutonium is unlikely to be returned in MOX or other forms any time soon. It is perhaps more likely that foreign stocks will remain in the UK indefinitely. Government policy reflects the uncertainty, concluding that foreign-owned plutonium could be managed alongside the UK's stock or 'title could be transferred to the UK, subject to acceptable commercial terms being agreed' (NDA, 2014, p.4). This would enable the UK to gain national control over more of the civil plutonium in the UK while also avoiding the need to physically transfer it elsewhere (DECC, 2015b).

The government's policy, then, is to repatriate the substantial stocks of foreign origin wastes, spent fuel and plutonium in some form or other. Transnational shipments of these materials in heavily guarded ships is highly contentious and has aroused considerable international protest especially at Cherbourg over shipments by sea from France to Japan, at points on the journey halfway around the world, notably, off the coast of Chile, and at the receiving port (Aubrey, 1993; Blowers and Lowry, 1996, see also Chapter 4). The annual rail transfers of vitrified wastes from La Hague to Gorleben in Germany were bitterly resisted (Chapter 5). There have been few shipments from the UK and, so far at least, protests have been relatively subdued. And, the prospect that these foreign nuclear materials will remain indefinitely to be managed in the UK likewise has not provoked noticeable public concern, for reasons I shall examine later in the chapter.

Sellafield – the nuclear dustbin

Shortly after THORP was finally commissioned in 1992, Sellafield became the focus of another controversy: the long-term management of the nation's radioactive wastes. Around two-thirds of the nation's nuclear wastes were already in varying conditions in so-called 'interim' stores scattered in buildings across the site pending their transfer to a permanent resting place. Although it might seem logical, inevitable even, that Sellafield would be the obvious location for the preferred solution of a deep geological disposal facility it did not feature in early efforts to find a suitable site. Nirex (the Nuclear Industry Radioactive Waste Executive), the organisation established to select a site, initially embarked on its quest elsewhere. In the fateful year of 1983 Nirex announced its selection of an abandoned anhydrite mine in Billingham on Teesside for a deep repository for long-lived ILW (intermediate level waste) and Elstow in Bedfordshire for a shallow burial facility for short-lived ILW and low level wastes (LLW).[3] It so happened that this LLW facility was needed because the existing disposal facility, at Drigg near Sellafield, was anticipated to be full by 1998. In fact, improvements in packaging and conditioning wastes, extensions at the site and use of other sites for very low level wastes have ensured that Drigg will remain open as the national repository for LLW until around the middle of this century.

Nirex's announcement of the sites immediately stirred the local communities into vigorous opposition based on a combination of fear of radioactive waste and outrage at being selected without any forewarning or consultation, a classic exercise by Nirex of the Decide Announce Defend (DAD) approach to site selection. Indeed, my own participation in radioactive waste policy and politics began as a county councillor opposed to the Elstow project (Blowers et al., 1991). Billingham was dropped after intense opposition in early 1995 while Elstow's oponents had to continue their campaign in conjunction with three other sites, all in Eastern England, chosen for a comparative assessment. Far from being able to divide and rule, Nirex encountered cross-cutting mobilisation of a united opposition at all four sites. The ensuing so-called 'Four Sites Saga' culminated in a provocative dénouement with barricades defending women and children and preventing hard-hatted contractors entering the sites. Such unfavourable images helped to persuade the government to accept the defeat of the Nirex proposals. While *The Times* in a leading article portentously described the protesters as 'Middle-class, middle-aged hooligans from middle England' (19 August 1986), their eventual victory might rather be described as a triumph of participative local democracy against arrogant and insensitive centralised decision making. The full story of the conflict is set out in my earlier book, *The International Politics of Nuclear Waste* (1991) written in conjuction with David Lowry and Barry Solomon.

From this defeat the government and Nirex recognised a new approach was needed. A fresh beginning was in prospect with the publication of *The Way Forward* in which Nirex declared its intention 'to promote public understanding of the issues involved and to stimulate comment which will assist Nirex in developing acceptable proposals' (Nirex, 1987, p.4). Nirex undertook an extensive comparative site selection process using multi-attribute decision making techniques to apply a

FIGURE 3.8 Protesters from near South Killingholme on the Humber Bridge

Source: Reproduced courtesy of Hull Daily Mail

FIGURE 3.9 Postcard protest against the Elstow site in Bedfordshire

Source: Bedfordshire County Council

range of scientific and economic criteria to various sites progressively narrowing down the field until, eventually, 'a picture emerged of three sites that could be recommended for further investigation, with possible merit in a further two' (Phillips, 1995, p.7), among them, Sellafield and Dounreay. Meanwhile, Nirex had also sponsored a public consultation exercise to determine reactions to the proposed disposal concepts and potential areas of search for sites (UEA, 1988). Unsurprisingly the upshot of this exercise was widespread public apprehension and unwillingness to accept radioactive waste. The consultation indicated only 'two areas where there is a measure of public support' and the then Secretary of State for the Environment, Nicholas Ridley, wrote to Nirex accepting that 'it would be best first to explore those sites where there is some measure of local support for civil nuclear activities'. To no one's surprise Nirex decided to focus further investigations on Sellafield (dropping Dounreay because of the high transport costs involved in such a remote location in the north of Scotland). Despite its conversion to a more consultative approach, the selection of Sellafield had all the hallmarks of a predetermined solution concocted through a closed process of expert decision making and relying on Sellafield as the path of least public resistance (Hinchliffe and Blowers, 2003). It was, in reality, a more sophisticated form of DAD. For Jean McSorley, 'The idea of Sellafield being the best place was laughable' (interview, 2014).

The flawed nature of the process became apparent in the planning inquiry lasting 66 days in 1995/6 into the proposal to construct an underground laboratory (grandly called a Rock Characterisation Facility, RCF) in order to explore the geological and hydrogeological features of the proposed site. The site chosen by Nirex on the basis of borehole drillings was close to Sellafield. Although the application by Nirex was for an underground laboratory it was recognised that the resources that would be sunk into the project amounted to making a 'pre-commitment' to the site becoming an underground geological repository should further investigation indicate a safety case could be demonstrated. The proposal was rejected by Cumbria County Council and Nirex appealed against the decision. As in the earlier inquiry over THORP, the RCF Inquiry tended to be an elite encounter where opponents were able to deploy counter-expertise to the scientific case put forward by Nirex. Although Friends of the Earth and Greenpeace played prominent roles, the chief protagonist was Cumbria County Council and there was a supporting cast of local interest groups. Once again Nirex was defeated and the RCF was rejected by the Minister on the advice of the Inquiry Inspector (Cumbria County Council, 1996). It may be observed that both Nirex defeats, on the shallow burial facility in Eastern England in 1987 and on the RCF in West Cumbria a decade later, were announced by the Minister on the eve of a national general election giving some emphasis to the political salience of radioactive waste. The RCF proposal was found wanting on three main counts: local environmental impacts; 'the scientific uncertainties and technical deficiencies in the proposals'; and 'the process of the selection of the site' (Government Office for the North West, 1997). In short, the proposal had been deficient both scientifically and socially; the scientific uncertainties and inadequacies of site selection were decisive in its rejection.

There had been little surprise in the eventual choice of Sellafield. It was widely believed, and the inquiry had provided the evidence, that the elaborate site selection undertaken by Nirex was a legitimation exercise conferring a rational justification on political expediency. For Sellafield, already host to much of the nation's waste, was the obvious, some might say the only, possible location for a rock laboratory *qua* deep geological repository. Yet, in this supposedly nuclear friendly and experienced community, Nirex had encountered well marshalled opposition spanning the county council, national NGOs and an array of local groups, spontaneously arising to fight the proposal. The RCF had been opposed for several reasons: it was not felt any benefits would compensate for the environmental detriment it would cause; the selection of Sellafield had not been adequately demonstrated or justified; and there were concerns that the geological circumstances of the area were unsatisfactory for the purpose. But, there were also cultural factors present in the community's opposition. There was the belief that the project was for a repository by any other name, an alien intrusion to take not only wastes already at Sellafield but those from other sites and other countries. This resentment was transformed among some into hostility against Nirex. Recollecting the inquiry years later, John Clarke, the Environmental Health Officer at Sellafield, commented, 'Nirex still exists, the name is still poison. It's a combination of suspicion and experience' (interview, 2004). A group of trade unionists I interviewed in 2006 reflected, 'Nirex is a bit of a swear word, people switch off. We are going to have to battle against that feeling'.

All the doubts about site selection and geology and the resentment and lack of trust that had typified the conflict over the RCF were to resurface fifteen years later when Sellafield once again was chosen, or rather was 'volunteered', as the possible location for a geological disposal facility. But, before that, new ideas, new policies and new processes would be developed in the hope of finding an acceptable solution for the problem of managing the nation's radioactive wastes.

A moment of transition

In several respects the defeat of the RCF proposal in 1997 marks a moment of transition. It was the point where the successive attempts to find a solution to waste management policy had run into the ground. The abandonment of the campaign to find suitable sites for high level waste disposal in 1981, the suspension of sea dumping in 1983, the defeat of Nirex, first in Eastern England in 1987 and now at Sellafield ten years later, left the government with nowhere else to turn for a solution. What was clear was that DAD had failed and the feeble attempt at securing public acceptability for the RCF had fared no better. In terms of finding a solution to the problem of wastes, the government and the nuclear industry had to start all over again, to rethink their approach.

Moreover, this was also a point that seemed to signal the demise of the nuclear industry itself in the UK. With the opening of THORP in 1994 and Sizewell B in 1995 the expansion of nuclear energy was over and the process of decommissioning the older nuclear power stations was about to begin. Although there continued to be flirtation with new projects to deal with the large plutonium stockpile at Sellafield such as another MOX plant, to all intents and purposes nuclear processing was winding down. In the early years of the new century it was becoming clear that Sellafield, like Hanford somewhat earlier (see Chapter 2), was facing a different future, a future committed to clean-up, decommissioning and waste management. The transition would prove to be painful, marked by a reluctance on the part of management and workers at Sellafield to come to terms with the new realities, an unwillingness to let go of reprocessing, anxiety about job loss but, ultimately, a recognition that the long-term future of Sellafield lay in cleaning up the legacy of the past. The transition was marked by organisational change with the state owned company, British Nuclear Fuels plc (BNFL), which had been associated with the reprocessing culture of Sellafield since 1971 but had become overextended and bankrupt, replaced by the Nuclear Decommissioning Authority in 2005 thereby putting the government's imprimatur on what had by then become the primary purpose of Sellafield. A consequence of the demise of BNFL was managerial fragmentation and privatisation which was later blamed for poor performance, cost escalation and inadequate leaderhip during the transition.

The transition in function and organisation was accompanied by a shift in the nuclear discourse. By the mid 1990s the discourse of Danger and Distrust had run its course and new circumstances, concepts and ideas were forming a new discourse. The nuclear industry appeared to be in terminal decline, brought on by its commercial weakness in the face of cheaper fossil fuels, notably North Sea gas, its own technical failings and accidents and by a public opinion fearful of the risks of nuclear energy and distrustful of its claims of safety and security. What remained was the problem of the long-term management of the country's voluminous legacy wastes. Every effort to find a solution to the problem of radioactive waste that was both credible and acceptable had failed. It was this problem that provided the political focus for the discourse that was now emerging.

The new discourse, which I have called the discourse of *Consensus and Cooperation* (see Chapter 1), achieved political momentum through the opportunities opened up by the decline of nuclear power and the need for a solution to the problem of its wastes. With no further development of nuclear energy or reprocessing in sight, political space opened up for the anti-nuclear movement to deploy its energies towards trying to solve the industry's most intractable problem. To a degree absent hitherto, both pro- and anti-nuclear interests had a common interest since the safe and secure long-term management of nuclear wastes was in everyone's interest. A more co-operative rhythm in nuclear politics was abroad, raising the prospect of some form of policy consensus. The changes in power relations were manifested in broader shifts in environmental politics and governance. This was the time of the 'deliberative turn' when environmental policy making was characterised by a commitment to openness, transparency, public and stakeholder engagement and notions of partnership and participative democracy. Radioactive waste management was at the epicentre of these new approaches and it led the way in developing innovative forms of participation and governance. This newly developing discourse of Consensus and Cooperation reached its apogee during the first decade of the new century. Once again, Sellafield was to find itself at the heart of the problem and the focus of the proposed solution to radioactive waste management.

Sellafield and the nation's nuclear legacy

A new way forward

In the aftermath of the RCF debacle, then, there was broad agreement among government, industry and environmental groups that a new way forward was needed. Moreover, there was agreement that a key reason for past failures was the absence of either scientific or social consensus on policy resulting from a lack of openness, trust and public acceptability. The search for solutions had to start again from scratch but in the knowledge that attempts to impose technical solutions through DAD would not work. At the turn of the century the mainsprings of a new way forward were being developed by three different bodies working on parallel but convergent lines.

One was the much maligned Nirex who, under new leadership, undertook a complete reappraisal of their approach placing emphasis on the new buzzwords of 'openness', 'transparency', 'participation' and 'partnership' to achieve consensus on radioactive waste management through a policy of community veto and volunteerism (Nirex, 2000; Atherton, 2000). Incidentally, as part of this new approach, I was invited to join the Nirex Board to help promote and develop this new sense of direction.

The second organisation was the government's scientific advisory body, the Radioactive Waste Management Advisory Committee (RWMAC) on which I also served. In a series of reports the RWMAC set out some guiding principles on which a more consensual approach to radioactive waste management could be based linking scientific and social criteria. RWMAC also stressed the idea of volunteerism with a deliberative process of engagement in decision making (RWMAC 1999, 2001). Finally, the House of Lords Select Committee on Science and Technology was likewise focusing on radioactive waste as a political and social as well as a scientific problem. While the Lords were committed to deep disposal as the only long-term option they, too, recognised the need for a 'volunteering approach' with a right of veto. They also saw the need for an independent group to oversee the process of siting a repository that 'is open, transparent, reasonable and robust' (House of Lords, 1999, p.50).

It seems remarkable how much consensus on the ways forward had been achieved by these various initiatives. Most notable was the commitment to openness and transparency which became the *leitmotif* of the discourse of Consensus and Cooperation. Overall, the need to secure trust and public acceptability became paramount objectives. There was agreement on the ultimate goal of policy: deep geological disposal as the long-term management solution. It was recognised that, for this to be achieved, it would be necessary to integrate scientific and social knowledge streams in order to develop a robust and implementable policy. Finally, there was an emerging agreement on *process*, the means by which geological disposal might be achieved. The main elements of this were: a phased programme; a volunteer process of site selection and veto; with policy formulation carried out by an overseeing body reporting to the Secretary of State. This developing consensus fructified in a major initiative put forward by the government. This was the so-called MRWS (Managing Radioactive Waste Safely) process.

Following a comprehensive consultation (Defra, 2001) the government set up a new, independent committee, the Committee on Radioactive Waste Management (CoRWM) (to which I was appointed) with the singular objective of recommending 'the options, or combination of options, that can provide a long-term solution, providing protection for people and the environment' (CoRWM, 2006, p.187). It was to work in an open and transparent manner to achieve recommendations which 'must inspire public confidence'. Its remit was to be confined to the UK's inventory of legacy wastes, basically the HLW and ILW already in store at Sellafield. At this point, no mention was made of wastes that might arise from new build.

For three years CoRWM set to its task with almost manic energy and enthusiasm. Essentially it developed four knowledge streams which were ultimately integrated to produce its recommendations. One, the scientific knowledge stream, was designed to identify the best options through an elaborate Multi-Criteria Decision Analysis (MCDA), laborious and time consuming but achieving the desired scientific credibility for CoRWM's work. Equally demanding was the extensive Public and Stakeholder Engagement (PSE) programme involving around 5,000 citizens in forums, open meetings, round tables, workshops and projects of various kinds. This amounted to a prodigious effort to distil the multifarious perspectives, values and preferences into a coherent, consensual set of 'common judgements on common interests founded on reason and argument' (O'Neil, 2001, p.491). Third was an ethical knowledge stream focused around concepts of fairness, of inter- and intra-generational equity, procedural fairness, responsibility and democratic decision making (CoRWM, 2007a). The fourth stream was knowledge of overseas experience. All the countries studied confirmed the scientific consensus favouring disposal as well as the necessity for any programme of waste management to take into account the need for social and political consent.

The various perspectives gleaned from these four knowledge streams were drawn together in a set of integrated recommendations opening with the pronouncement:

> Within the present state of knowledge, CoRWM considers geological disposal to be the best available approach for the long-term management of all the material categorised as waste in the CoRWM inventory when compared with the risks associated with other methods of management. The aim should be to progress to disposal as soon as practicable, consistent with developing and maintaining public and stakeholder confidence.
>
> *(CoRWM, 2006, pp.12–13)*

There was an important qualifier that 'A robust programme of interim storage must play an integral part in the long-term management strategy' (p.13). In addition the Committee recommended a commitment to research and development into the long-term safety of geological disposal and that the 'possibility that other long-term management options' might emerge should be left open. Overall, the Committee considered it had met its basic aim of producing recommendations that commanded public confidence while producing a set of interrelated proposals that the government could adopt and implement.

As for implementation, CoRWM produced a further set of recommendations drawing on earlier work of Nirex, RWMAC and the House of Lords as well as overseas experience. CoRWM's implementation report was a paean to all the principles of the discourse of Consensus and Cooperation covering openness, transparency, deliberation, ethics, integration of knowledge and democratic decision making. CoRWM set out its own approach based on three concepts, the three Ps of Participation, Partnership and Packages (CoRWM, 2006, 2007b, 2007c), each of them the focus of specific recommendations for a process for siting a repository. First, in terms of participation, community involvement 'should be based on the principle of volunteerism, that is, an expressed willingness to participate' (2006, p.14). Second, involvement should be achieved 'through the development of a partnership approach, based on an equal and open relationship between potential host communities and those responsible for implementation'. And, third, willingness to participate should 'be supported by the provision of community packages that are designed both to facilitate participation in the short-term and to ensure that a radioactive waste facility is acceptable to the host community in the long-term' (ibid.). It remained for this elaborate and elegant, though conceptual, set of implementation proposals to be transformed into a working practice en route both to develop a repository concept and to find an appropriate site for its underground location. This required both a government commitment and a willing community which is where, once again, Sellafield was called upon.

Government takes the plunge

CoRWM's recommendations were quickly seized upon by the government in its response as offering a 'sound basis for moving forward' (Defra, 2006, p.3). But, even at this early stage there were a few intimations of differences of emphasis between CoRWM's ideas and the government's intentions. In particular, it soon became clear that the government was focused on the idea of geological disposal as quickly as possible as a means of demonstrating that radioactive waste would not be a barrier to a new nuclear programme on which it was about to embark. Consequently, the more qualified and long-term strategy of integrated storage and disposal proffered by CoRWM receded as the government proclaimed its intention 'to move forward as fast as is practicable to develop a strategy for the delivery of geological disposal, in a manner that is scientifically sound, develops and maintains public confidence, and ensures the effective use of public monies' (ibid., p.6).

As the possibility of new build, styled a 'nuclear renaissance', began to take hold of the government's imagination as the solution to the twin problems of energy and environmental security – more popularly styled as 'keeping the lights on' and 'saving the planet' – so CoRWM felt its work should not be co-opted into the government's new policy initiative. Its remit was to deal with the nation's burden of legacy wastes, not those that might arise from a fleet of new power stations. CoRWM was emphatic that new build wastes should be subject to a separate process since the size and nature of the inventory raised potentially different technical issues in terms of design, size and number of possible repositories. In short, there was 'a range of issues including the social, political and ethical issues of a deliberate decision to create new nuclear wastes' (CoRWM, 2006, p.14). Therefore, CoRWM's proposals were not 'to be seized upon as a green light for new build' (p.15). Nevertheless, the government, in its consultation on radioactive waste management policy, was clear that 'in the event that there were new nuclear power stations, waste and spent fuel from those stations could be accommodated in the same geological disposal facility' (Defra, 2007, p.12).

The underlying presumption of the CoRWM process was that a geologically suitable site could be identified through a voluntary process. The CoRWM process essentially envisaged a twin track approach to identifying a site which was at once scientifically sound and publicly acceptable. Experience in other countries – Sweden, Finland especially – were cited as using a similar approach. But, there was a crucial difference. In the Nordic case sites were identified *before* the communities were invited to volunteer, in the UK process sites themselves would be identified voluntarily. It would be left open for volunteers to come forward. As critics argued at the time, this could lead to a situation where a publicly acceptable site was not necessarily satisfactory from a geological point of view. It was an issue of voluntarism before geology, not, as might seem more logical, geology before voluntarism. This would become the critical issue when the theoretical debate moved to the territorial context of West Cumbria.

Another area of potential difficulty lay in defining the decision making bodies. These would be the local authorities, democratically elected bodies with decision making powers, which would take the key decisions such as entering the siting process, exercising the right of withdrawal and taking the decision to proceed. Where there was only one level of local government, as in the unitary authorities, this raised no problem of responsibility. But where two tiers of local government exercised power over the same territory, as was the case in West Cumbria, there was potential for disagreement and impasse.

The government confirmed its approach to implementing geological disposal in a White Paper in June 2008 (Defra, 2008a). All the features of the 'deliberative turn' (see Chapter 1) were there – the emphasis on a voluntary process, an open and transparent approach, engagement with stakeholders and public, development of partnerships and a commitment to devolved democratic decision making at local government level. And there is no reason to think that this commitment was anything less than genuine. For the present, the government was content to give voluntarism its head. The White Paper was a tribute to how far the deliberative turn fostered by a discourse of Consensus and Co-operation had penetrated government. By contrast with the Managing Radioactive Waste Safely process (MRWS) as the approach developed by CoRWM and adopted by government came to be called, the contemporary consultations on new build seemed to revert to a more conventional decide–consult–declare approach to decision making. It was a more formalised, routine set of consultations on developing and siting new nuclear reactors at eight potentially suitable sites, one of which, incidentally, was on land adjacent to Sellafield, which later became known as Moorside.

In general, there was support for the government's framework for implementing geological disposal (Defra, 2008b). The main criticisms that emerged concerned those areas where the government had deviated from CoRWM's core recommendations, for example, by placing emphasis on geological disposal as soon as practicable rather than CoRWM's more measured integrated process of storage and disposal. There was also the worry that 'if only volunteer sites are considered, more promising sites, geologically speaking, might be missed' (p.9). The potential for sabotaging CoRWM's work was recognised by those who 'held the view that the apparent haste to decide on a new generation of nuclear power plants is undermining policy development on radioactive waste and the groundwork done by CoRWM' (p.31). There was also the question of what might happen if the process failed. 'If no one volunteers, or if volunteer sites do not have "suitable geology", what would be the next steps?' (p.9). The government could not say but nevertheless had an answer. 'In the event that at some point in the future, voluntarism and partnership does not look likely to work government reserves the right to explore other approaches' (Defra, 2008a, p.47). There is, thus, more than a hint that the real purpose of radioactive waste management policy was to find a quick fix to help legitimate new build.

Sellafield – an inevitable expression of interest

The government soon put its White Paper proposals into action by issuing an invitation to communities 'to express an interest in opening up without commitment discussions on the possibility of hosting a geological disposal facility at some point in the future' (Defra, 2008a, p.69). The invitation was deliberately vague in terms of definition of community and it was open ended in the sense that the option to express an interest 'will be left open for the foreseeable future' (ibid., p.69). But, it was also pretty clear where the government expected the earliest response to come from since 'it is expected that some communities may be better informed on the issues than others, for example, those who already have local nuclear facilities' (ibid., p.69). Above all, it seemed almost inevitable that the area where two-thirds of the nation's wastes were already in store would be the first to come forward.

And so it proved. Even before the starting gun was officially fired, Copeland Borough Council, which covers the Sellafield nuclear complex, made its intentions clear. The council leader, Elaine Woodburn, said the community would ask for an 'endowment' from the government since the repository, if it could be safely located in West Cumbria, would 'be here for thousands of years. We can't ask for projects that will last 50 or 100 years because that would be a disservice to future generations' (*The Observer*, 8 June 2010). Copeland's almost precipitate action was quickly followed by neighbouring Allerdale Borough Council, much of it within commuting distance of Sellafield, with its own expression of interest. Not to be outdone in staking a claim, Cumbria County Council, the upper tier local authority covering a wider region, including the Lake District, Eden Valley, Carlisle and the Furness area around Barrow-in-Furness as well as West Cumbria, also joined in talks. So, by early 2009, all three councils had taken the first tentative steps in making formal expressions of interest in the GDF (Geological Disposal Facility) siting process. Almost immediately, it seemed, the government had landed its catch, in the very heartland of the nuclear industry where the lure of jobs and investment to sustain the economy could not be ignored.

Yet, this 'expression of interest' was only a cautious, guarded and qualified first step, a tentative '"without commitment" interest in discussions about potential involvement in the siting process' (Defra, 2008a, p.49). The underlying strategy was a softly, softly approach that might initially prove tempting leading eventually to an ever tightening embrace as the communities became familiar with the prospect of hosting the nation's nuclear wastes. At this early stage, too, the possibilities of compensation for willing communities, the 'Packages' element of CoRWM's three Ps of implementation, were not overlooked with the Secretary of State, Hilary Benn, declaring that it was 'only fair to reward them for this essential service to the nation' (*Independent on Sunday*, 29 June 2008). Although in West Cumbria the show was on the road, any early optimism about a smooth passage to finding a site for a repository would soon be qualified. For one reason, the apparent enthusiasm of the councils was not universally experienced in the region and, as time went on, conflicts developed. For another, only West Cumbria was being volunteered; elsewhere across the country there was silence.

The chosen vehicle of the voluntarist approach was the West Cumbria Managing Radioactive Waste Safely Partnership (WCMRWS) comprising a range of representative interests including the three councils who had expressed interest, the Lake District National Park Authority and Cumbria Association of Local Councils (CALC), voluntary bodies and business interests (churches, Chambers of Commerce, trade unions, farming, tourism) and the Nuclear Legacy Advisory Forum (NuLeAF) representing local authorities concerned about nuclear legacy issues. Altogether there were 29 members, over half of them elected members of the three councils. The composition and balance of the Partnership was to prove a key issue in the decision making process and its final outcome. Noticeably absent were any environmental NGOs, who had misgivings about the process and concerns about geological disposal and declined the invitation to occupy two seats on the Partnership preferring the freedom to criticise from without rather than participate from within. They conducted an external campaign raising public awarenesss but also attended the public sessions of the Partnership and contributed

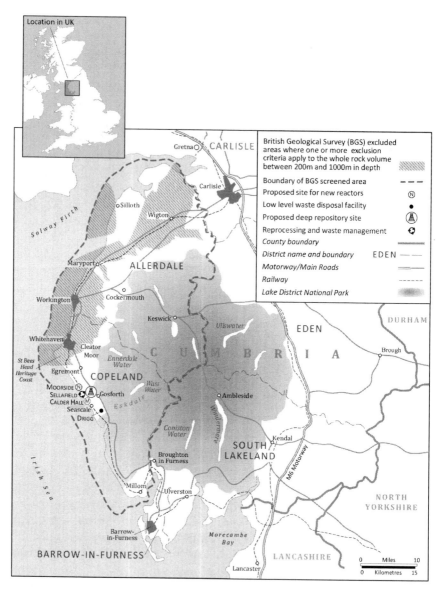

FIGURE 3.10 Map of Cumbria, northwest England

Map by: John Hunt

responses to consultations. Their activity in arousing public anxieties, promoting alternative arguments and possibilities as well as challenging the Partnership and the local councils had an important, possibly crucial, impact on the eventual outcome of the process.

For a period of three years (2009–12) the Partnership undertook a prodigious amount of work, in deepening its understanding of the issues and in raising public awareness and generating involvement through an energetic and comprehensive series of public and stakeholder engagement programmes. Like CoRWM before it, the West Cumbria Partnership proved a model of participative democracy in practice operating within the discursive theory and practice of cooperation and consensus. However, the effort to achieve consensus could not conceal very real differences and antagonisms operating both within and without. By far the most significant of these was the debate over geology.

Geology and/or voluntarism?

Put simply, was there anywhere in West Cumbria where the geological, or more pertinently the hydrogeological, conditions were capable of providing a safe tomb at depth for the over one million cubic metres cavern covering between six and eleven times the size of the Albert Hall (West Cumbria MRWS Partnership, 2011a)? The geology of Cumbria is highly complex yet it is one of the most investigated regions of Great Britain. During the borehole investigations for the RCF in the 1990s, Nirex had concluded that the area held 'good promise as a suitable location for a repository' (Nirex, 1993, p.40) though it had also conceded that 'there are likely to be radiologically better sites available around the UK' (Government Office for the North West, 1997, section 7). Indeed, the Head of Science at Nirex at the time confessed there were problems at the site. 'I have the feeling we may struggle to make a case for the site' (Nirex, Internal Memo, 10 December 1996). Thus, even back then there were doubts about its geological suitability.

The doubts were presented in a detailed compendium of submissions to the RCF Inquiry compiled by two geologists Stuart Haszeldine and David Smythe (1996a, 1996b), then at Glasgow University and who later re-emerged as prominent and indefatigable opponents of the GDF in West Cumbria. With the Partnership established, the British Geological Survey (BGS) carried out a broad screening exercise which eliminated an area to the north and west of West Cumbria as unsuitable (BGS, 2010); within the remaining area more detailed criteria would be used to identify if there were any areas of promise (Figure 3.9). Now the geological battle lines were drawn. On one side the NDA, the body responsible for managing wastes, took the view that there were areas potentially suitable within West Cumbria. On the other side the two professors Haszeldine and Smythe tenaciously claimed that scientifically all of West Cumbria was unsuitable. This assertion was challenged by the Chair of CoRWM who stated 'there is presently no credible scientific case to support the contention that all of West Cumbria is geologically unsuitable' (letter from Prof. Robert Pickard to Elaine Woodburn, 16 February, 2011). The Partnership's independent expert, Dr. Jeremy Dearlove, argued that the possibility of finding suitable geology was sufficient to justify further intensive in-depth research. The debate became personal, polarised and politicised with Smythe and Dearlove engaged in increasingly vituperative counteraccusations. In one exchange Dearlove accused Smythe of peddling his 'personal opinion and not the opinion of the wider geological community, that the entire MRWS Partnership area is geologically unsuitable' (FWS Consultants, 2011). In his turn, Smythe accused Dearlove of trying 'to represent me as a lone maverick earth scientist pitting his solo opinions against a supposed "*wider geological community*"' (Smythe, 2011). It was acknowledged that the geology was complex and the possibilities had, over the course of the debate, been narrowed down to two relatively small areas, in the north near the Solway Firth (Silloth) and in the south in the area of Eskdale. Even these were disputed. Nonetheless, the Partnership eventually concluded, albeit tentatively, that 'the absence of clear, detailed evidence that demonstrates that all of West Cumbria should be ruled out has led us to the initial opinion that there is enough possibly suitable land to make further progress worthwhile' (West Cumbria MRWS Partnership, 2011a, p.34).

The nub of the issue was whether geology alone should be sufficient to ensure the wastes were contained or whether geology plus engineered barriers could be effective enough. This question became posed as a choice between the 'best' available site chosen on scientific grounds or a site that was volunteered where the geology (perhaps with engineered barriers) though not necessarily the best was nevertheless quite acceptable in the sense of being able to meet the safety case. This conflict between geology first and geology plus voluntarism became crucial at the climax of the debate in West Cumbria. What the debate had undoubtedly established was that, in West Cumbria at least, geology was as critical in the process of site selection as the voluntary principle. In the view of David Smythe and others the placing of voluntarism ahead of geological suitability was unjustified. It was perhaps not a case of voluntarism or geology, rather a case of geology and voluntarism for, in the case of West Cumbria, it became clear that the two were not mutually exclusive but mutually entangled in what was becoming an increasingly politicised dispute.

Consensus and conflict

At the outset of the MRWS process in West Cumbria the sense of a community working in partnership was, at least rhetorically, quite high. Elaine Woodburn, Copeland's Leader, stressed, 'We want everyone to have the chance to help us decide', a sentiment echoed by Tim Knowles, a leading county councillor also from Copeland, 'We want to be completely satisfied that locating the facility will be in the best interests of the area and is supported by local people' (*Whitehaven News*, 12 March 2010). But, once the work had begun in earnest the inherent political and geographical tensions began to emerge – between elected representatives and other stakeholders, between the different levels of local government (county and district) and between West Cumbria and other parts of the county. The differences, at first rather vague but increasingly more specific, developed as the Partnership delved more deeply into the various issues – geology, inventory, the scale of the repository, new build wastes, the timescales for disposal and the ongoing problem of dealing with storage of wastes at Sellafield. A political cleavage was developing in the community at large, or at least among those inclined to participate in the debate. And the dividing line was not over the rather tentative question the Partnership was pondering, whether the community should take part in the search for a site in West Cumbria, but, rather, the more direct variant, whether the project should be rejected, at least in West Cumbria. Backing for the possibility of the repository came from predictable sources supportive of the nuclear industry and especially Sellafield: the trade unions, Copeland's MP and its council leadership. On the other side were anti-nuclear groups, some long established like Friends of the Earth and CORE together with newly emerged campaigning groups like Radiation Free Lakeland, Save Our Lake District – Don't Dump Cumbria (led by local Friends of the Earth campaigners) and evanescent groups defending specific local communities such as Solway Plain Against Nuclear Dump (SPAND). For the first time in the history of conflicts over the nuclear industry in West Cumbria there was a high degree of mobilisation by local groups able to focus their campaign on local decision makers in the Partnership and on the councils. Unlike previous major decisions such as THORP or the RCF which were taken by government, the decision whether to proceed with the GDF had been devolved thereby releasing local energies on local decisions.

Over the period of the Partnership an intensive effort at engagement with the public was undertaken. All in all, as an exercise in public participation, the West Cumbrian effort must rank as one of the most comprehensive, disinterested, sophisticated, open and interactive ever undertaken. As an exercise in public engagement it ranks with CoRWM's pioneering approach. According to a poll with a response rate of just under half taken during the engagement process just over half those responding declared they were aware of the search (74% Copeland, 65% Allerdale, 49% Cumbria). Over half (53%) supported continuing to take part in the search with a third (33%) against and figures broken down for Allerdale and Cumbria county were similar while Copeland showed more than two-thirds (68%) in favour and only a quarter (23%) against, as might be anticipated in the area surrounding Sellafield (Ipsos MORI, 2012). Of course, such a poll provides only a superficial impression and needs to be set against the evidence of responses to the consultation indicating 430 in favour of proceeding with 320 against although little store was set by this crude numerical yardstick (West Cumbria MRWS Partnership, 2012a). Meanwhile, the Cumbria Association of Local Councils (CALC) had found that only 8 of 88 town and parish councils supported proceeding further while 43 were against with 37 not expressing a position. However, most of the parishes 'raised issues of one kind or another with 70% not supporting and 25% supporting proceeding to the next stage' (CALC, 2012). CALC, though not recognised as a decision making body, expected their views to be given weight as elected community representatives arguing that any decisions taken would not be credible unless supported by the parish tier of local government. In their judgement, taken together, surveys suggested opinion was evenly divided and there was no clear evidence of public support sufficient to move the process forward.

A basic concern among the public was the focus on disposal as a take it or leave it option without alternatives and that West Cumbria was the only area under consideration as if the government was hell bent on finding its desired solution in its preferred location as quickly as possible. The Partnership got the point: 'We consider that many of the issues raised ... arise, rather like the tips of icebergs, from more fundamental questions of trust, especially in future decisions, decision-makers and organisational arrangements' (West Cumbria MRWS Partnership, 2012a, p.122). The uncertainties and divisions within the community were reflected in the final rather tentative conclusions of the Partnership. 'Our overall opinion is that, at this stage, we are fairly confident that an acceptable process can be put in place to assess and mitigate negative impacts, and maximise positive impacts' (West Cumbria MRWS Partnership, 2012b, p.6). Throughout its work the Partnership had striven for consensus accommodating the full range of views. But, 'where that has not been possible we have laid out the source of disagreement and explained the range of views present within the Partnership' (ibid., p.13). At the last, the Partnership reneged on its earlier commitment to make recommendations and left it to the councils to decide, lamely pronouncing:

> We, as the Partnership, feel it is important for the councils to be able to weigh up our work and opinions across the range of topics and issues ... before making a decision, and, as such, we have not provided a final recommendation about participation in the next stage of the MRWS process.
>
> *(Ibid., pp.12–13)*

This was a critical non-decision for it indicated uncertainty and a failure on the part of the Partnership to try to bring its work to a purposive conclusion committing its members to a specific outcome. Instead, the three councils who had been the dominant force within the Partnership were, as the decision making bodies, given unrestrained freedom to make up their own minds.

This freedom was given a more exquisite twist by a Memorandum of Understanding (West Cumbria MRWS Partnership, 2012c) whereby the councils agreed they would take the decision whether or not to proceed simultaneously and all would need to agree. In the event that one or more of the three authorities took a different position then the process would be paused. The government accepted that in the event of disagreement, 'the principles of partnership to which we have all been committed cannot be met. Accordingly, we would not proceed with the Managing Radioactive Waste Process in West Cumbria' (West Cumbria MRWS Partnership, 2011b). Prophetic words indeed! Although the three councils were each supportive of the nuclear industry in West Cumbria and were in favour of the prospect of new nuclear build at Sellafield there were also potentially differences in regard to their roles and perspectives on the GDF issue. While the two boroughs were focused on the possibility of hosting the GDF, the county council took a wider view that a GDF would be likely to impact upon other areas within the county. The potential grounds for dissent among the councils were inherent in the political geography and the Memorandum of Understanding provided the protocol for division.

Dénouement

With the Final Report published it was left to the councils to make up their minds on the question of whether to enter Stage 4. The distinction between Stage 3, expressing an interest, and stage 4, deciding to participate in a siting process, was not merely semantic; it was substantive. It was widely perceived as the moment when West Cumbria would become tied in, perhaps irrevocably, to a process that would lead inevitably to the GDF, notwithstanding the right of withdrawal. The issue now became thoroughly politicised as those opposed to further participation mobilised resources of protest to bring pressure on the councils.

The lack of sufficient knowledge and time to take a well-founded decision became the lynchpin of the opposition's case. A month before the decision, public meetings were held addressed by three professors, the two geologists, Stuart Haszeldine and David Smythe and myself, to give a social and political analysis of the situation. The geologists reiterated their argument that Cumbria's geology was complex, there were upward flows of groundwater and the two areas that had been identified as possibilities must be ruled out on grounds of permeability, faulting and topography. I argued that a decision to proceed would be premature in view of questions about geology and inventory, and the lack of comparative strategies or alternative areas of search for sites. There was a need to take the decision in a measured way based on a clearer affirmation of community support, not to succumb to pressure exerted by government intent on a dash for geological disposal as soon as possible. I concluded that 'if there are doubts on any of these issues it may be prudent not to proceed to Stage 4 at this time'. And so it came to pass, a few days before decision day set for 11 October 2012, the three council leaders decided to defer the decision for three months and ask for more time to get further clarification and information 'on a number of issues they believe are key to the issue of trust'.

This issue of trust began to dominate the twilight months of the MRWS process in West Cumbria. The Final Report had mused on this: 'A lack of trust appears to us to be at the root of many of the key concerns raised by the public and stakeholders' (West Cumbria MRWS Partnership, 2012b, p.54). The councils demanded reassurance from government on three major issues: the demand for a legally binding right of withdrawal; clarification of the government's commitment to community benefits; and, most crucially, given the uncertainty about Cumbria's geological suitability, they argued that alternative radioactive waste management solutions should be considered in parallel with the MRWS programme' (letter, three leaders to Baroness Verma, Minister for Environment, 1 October). There were doubts that these major issues could be sufficiently clarified in time for the deferred decision day, fixed for 30 January 2013.

As the day approached, groups opposed to the process continuing in Cumbria turned up the volume of protest with public meetings, petitions, demonstrations and press and media publicity, websites and social media campaigns. The more established groups (Friends of the Earth local groups, together with Save Our Lake District, CORE and Radiation Free Lakeland) were joined by newly formed

FIGURE 3.11 Protesters against the proposed geological disposal facility in West Cumbria

Source: Irene Sanderson North Cumbria CND

protest groups focusing on the areas identified as geological possibilities, SPAND (Solway Plain Against Nuclear Dump) and NoEND (No Ennerdale Nuclear Dump). In this pervid atmosphere the pressure was on the council leaderships, especially Cumbria County Council, regarded as the one most likely to say no to further participation. The final result brought its own drama with Allerdale and Copeland voting 5–2 and 6–1 respectively in favour but Cumbria, unpredictable to the last, voting by 7 to 3 against further participation thereby bringing the MRWS process in West Cumbria to an end.

The speech by Eddie Martin, Conservative Leader of Cumbria, argued the case for withdrawal. Government assurances on the right to withdrawal were not enough, attention should be focused on 'the safest place, not the easiest' and there was the fear that 'if the area becomes known in the national conscience as the place where nuclear waste is stored underground, the Lake District's reputation may not be so resilient' (*North West Evening Mail*, 30 January 2013). He had come to the conclusion that alternative solutions must be considered and, in Cumbria, storage (with considerably greater investment) was the way forward. In an impassioned peroration he set out his belief that it was the right decision:

> May I suggest, Members, that we put aside the politics and the science and the speculation, and the scaremongering … and trust the people, but …

well-informed people. Let's embrace the opportunity we now have; take the heat that will no doubt be generated by our decision and make the hard and difficult decision, knowing that we are doing it to make things better, not worse, for the majority of the people and the children and the future children of Cumbria.

On the morrow of the decision, the Secretary of State for Environment and Climate Change, Ed Davey, recognised that, following Cumbria's vote, 'the current process will be brought to a close in West Cumbria' (Press Release, Energy Secretary responds to Cumbria nuclear waste vote, DECC, 30 January 2013). Although disappointed, he reaffirmed the government's commitment to new nuclear power stations and to a policy of deep disposal based on voluntarism and a community-led approach. After time for reflection, the government would have to begin again.

So, the most committed nuclear community in the country with by far the greatest accumulation of radioactive wastes in stores had rejected the chance of a deep disposal facility in which those wastes might, one day, be buried. Along with earlier decisions on THORP and on the RCF, the decision not to proceed with the GDF was grounded in the social, economic and political geography of West Cumbria. In the next section I shall explore the developing and dynamic relationship between Sellafield and the local community as the nuclear discourse has changed over the years. I shall seek to show how the peripheral nature of West Cumbria explains the paradox of power and powerlessness that is at the heart of that relationship.

Sellafield – a peripheral paradox

Remoteness – geography and community in West Cumbria

Sellafield was chosen as the site for the production of plutonium for Britain's atomic bomb for a variety of pragmatic reasons. According to Harold Bolter (1996) it was not the first site to be considered. Among the early options rejected were North Wales (too attractive to tourists) and Arisaig in remote northwest Scotland (site conditions 'totally unsuitable'). Remoteness was an important criterion although nowhere in the country could the locational characteristics, notably the extensive remoteness that had been applied to Hanford, be found (see Chapter 2). While the technology developed for the first British reprocessing plant 'needed a relatively remote site, it did not need an Arisaig' (ibid., p.36). The post-war planners eventually settled on Sellafield, on the site of a wartime Royal Ordnance Factory supplying TNT for armaments. Its remoteness from population centres offered the advantages of relative safety, security and secrecy together in an area of declining industry with a small local workforce familiar with rather clandestine and risky operations. 'In all, Sellafield was one of the few areas of the country which would welcome the new industry' (ibid., p.36).

This new industry represented a modern technology inserted into a traditional rural and industrial landscape. Agriculture, mainly pastoral small farms of dairy herds and sheep and fishing, was the basis of the rural economy and close-knit, territorial society described by Williams (1956) in his post-war study of Gosforth, a village close to Sellafield. He was aware of the changes already well in train 'as a result of the increasing influence of urban culture' (Bell and Newby, 1971, p.146) and, writing in the early 1950s, perceives that the developments at Windscale brought in people 'who do not "fit in" and who have urban values' (ibid., p.146). But, soon these people had transformed both the landscape and social relations in the area, introducing a substantial industry whose influence spread throughout West Cumbria. In the immediate vicinity, villages such as Gosforth attracted the scientists and managers while Seascale became a 'works village' built to house workers and almost entirely dependent on Sellafield for its livelihood (Macgill, 1987, p.68).

Further up the coast are the towns of Whitehaven and Workington, each with a population of around 25,000, with the smaller town of Maryport to the north. Situated on the West Cumbrian coalfield and close to iron ore supplies in the Furness district to the south, they developed flourishing iron and steel industries in the 19th century and other manufacturing, notably chemicals. During the 20th century West Cumbria was an area in economic decline, a sub-region of uneven development. The prosperity derived from investment and good wages at Sellafield is barely evident in the declining towns, impoverished housing estates and neglected social infrastructure scattered around the region. The juxtaposition of well-being and poverty is evident in the vivid contrast between St Bees village at the most westerly point on the coast named in 2014 as the best place in Britain to raise a family and nearby Mirehouse and Woodhouse estates on the southern edge of Whitehaven which have high concentrations of multiple deprivation. The inequalities within West Cumbria contribute to its sense of difference and division but, at the same time, distinctiveness and isolation.

This sense of isolation is reinforced by the area's geographical insularity. To the west is the Irish Sea, to the north the Solway Plain, to the south the Furness district leading towards Barrow-in-Furness, itself a town dependent on production of nuclear submarines but a distinct sub-region. On its eastern side, West Cumbria is bounded by the Lake District, England's most renowned landscape and tourist honeypot. The boundary of the Lake District National Park is a geographical, economic, cultural and social frontier. The perception of the distinctiveness of West Cumbria was summed up in a sociological study of the impact of Sellafield in 1993:

> West Cumbrians evidently saw their area as 'different' and separate from the rest of society. This was expressed often as physical isolation and lack of communication with the rest of the Lake District even, let alone the rest of the country, and especially about the 'run-down and decrepit' appearance of the area, and about the lack of roads and mobility from the Sellafield vicinity.
>
> *(Wynne et al., 1993)*

FIGURE 3.12 Ennerdale Water, Lake District National Park

Source: therealgaffney.zenfolio.com

FIGURE 3.13 Woodhouse housing estate, Whitehaven

Source: BBC Radio Cumbria

The combination of insularity, inequality and integrity that characterises West Cumbria goes some way to explaining both its tenacity and ambivalence with respect to the nuclear industry. As Adrian Simper, Director of Strategy and Technology at the NDA, put it to me during an interview in 2014: 'The people of West Cumbria are nuclear savvy. Nuclear is unique. Geographical isolation breeds insularity, less economic flow in and out is usual'. The nuclear industry has meant investment and jobs in an area that was deindustrialising and with no evident signs of revival. Yet, the industry's influence, at once pervasive and fragmented, has brought benefits to a dependent workforce but has barely enhanced the well-being of large parts of the population living in backward rural areas or depressing urban environments. Sellafield, for all its divisiveness, is nonetheless all they have by way of major industrial activity and it has long commanded the loyalty of its workforce and the support, albeit sometimes reluctant, of the wider community. This support is rooted in the early years of development when the discourse of Trust in Technology was near universal in the country and Sellafield was its apotheosis. In later years, during the period of the Discourse of Danger and Distrust when THORP became a focus of controversy, support within the community remained firm. And, the nuclear industry has continued to draw on the reservoir of support within West Cumbria during its flirtations with reprocessing, MOX and other unsuccessful ventures as well as its commitment to the projected new nuclear reactors at Moorside separate from but adjoining Sellafield. The geographical isolation and boundedness of West Cumbria has bequeathed a defensive and largely unconditional support for its nuclear industry.

However, the support is not entirely unquestioning; rather it is a combination of various attitudes, positive and negative. Sally Macgill, in her 1987 study of Sellafield and the cancer controversy, summarised this complexity of attitudes to Sellafield thus:

> To some Sellafield is a unifying force of community life and self-confidence; to others a pale shadow of its former glory at the forefront of nuclear technology anywhere in the world; to others no more than a workplace; and to others a source of doubt, concern, or dread over feared radiation contamination effects.
>
> *(Macgill, 1987, p.7)*

As we have seen, community support for the nuclear industry seemingly does not extend to the idea of a deep disposal site for the long-term management of radioactive wastes in West Cumbria. On both occasions when this project was being promulgated, by Nirex with the RCF in the 1990s, and by the government with the GDF during 2009–12, there was considerable opposition aroused within West Cumbria and beyond. Part of the explanation for this lies, once again, in the geographical remoteness of West Cumbria in relationship to the rest of the country and the values and loyalties this promotes.

The remoteness of West Cumbria is frequently observed as a reason for the nuclear industry's presence and the absence of any significant development that is not nuclear dependent. 'We look peripheral on the map – it's really the peninsula effect. Cumbria is a Polo mint' (John Grainger, interview 2005, referring to Polo, the mint with the hole). Certainly, in relative terms, West Cumbria is a long way from other urban centres, around two hours (70 miles) to reach the motorway and a further four hours to reach London. Sellafield itself is served by a tortuous railway line leading north to Carlisle or south by way of Barrow to Lancaster. There is no airport and communications within West Cumbria are poor. 'People will not come because of transport infrastructure' (Keith Hitchen, interview, 2014). 'We're up here at the top left hand side of the country and don't get any of the facilities we need' (Tim Knowles, interview, 2014).

Geographical remoteness intensifies a feeling of neglect that breeds frustration and resentment especially towards outsiders. One of the reasons Nirex was opposed so vigorously over the RCF was that the company appeared as an intruder to be treated with suspicion. Sellafield was altogether different. The contrast was well summed up by Rex Toft, former Leader of Cumbria County Council, whom I met in 2004:

> This is a company area. People in Copeland live and work in Sellafield. BNFL pays good salaries with good pension schemes. That doesn't say we're beggars or that we can be bribed. If there were a committee set up to watch the tide come in, Nirex would give them money.

Nirex appeared as an arrogant outsider unsympathetic to local interests. 'Nirex illustrates the context of doing it wrong. What we don't want is another Nirex. Nirex didn't involve us in any way and ignored geology' (Geoffrey Blackwell, interview, 2004). Likewise, there was an aversion to the GDF fuelled by the concerns about geological suitability and burying wastes deep underground for perpetuity. Storage of the nation's legacy wastes already at Sellafield was one thing, permanent disposal including wastes from new build was quite another. Martin Forwood of CORE made the point: 'It would be ludicrous to me to move it from Sellafield given the risks of transport. It would be absolutely ridiculous. But Sellafield shouldn't necessarily be taking more' (interview, 2004).

The feelings of being embattled, different, put upon and disdained all have a geographical origin in the physical remoteness of the West Cumbria region. The idea of 'Us and Them' comes over strongly in this quote from a resident of Seascale recorded by Wynne et al.

> Well I think you know, when they haven't lived here. I mean I think the views of other people coming in – I mean even if you go away on holiday and things like that and they ask you where you live, and as soon as you mention Sellafield, you know there's a big reaction … as if you glow in the dark or something like that …

(Wynne et al., 1993, p.43)

The sense of being on the periphery, distanced and distinctive from other areas, is also related to the economic relationship between Sellafield and West Cumbria.

Economic dominance and dependence

The most striking and abiding feature of West Cumbria's economy is Sellafield's pre-eminence in employment and investment. Over the years its significance as the dominant employer has grown as the area's traditional manufacturing industries have declined and disappeared. Sellafield, too, has in recent times, moved from a predominantly manufacturing (reprocessing) role to one of clean-up and waste management, a transition that is in sight though not yet complete. During its history, as Adrian Simper recalls, the Sellafield site has responded to different calls, first to the call for making plutonium for atomic bombs, then to produce electricity from nuclear energy, followed by a call to produce fuels through reprocessing for new reactor designs, a 'time of nuclear nirvana and fast reactors'. Then the production slowed down, nuclear was unpopular and Sellafield was no longer at the forefront of nuclear technology but clearing up the mess created in its heyday. Simper comments, 'It is very difficult for the site which has already responded three times. Sellafield and the UK are at an early stage in transition. Hanford and Dounreay (since 1994) are further along. Sellafield reaches the equivalent stage by 2030. It's hard to move the mind set while still reprocessing' (interview, 2014). As at Hanford, so at Sellafield there has been a tendency among the workforce to cling

FIGURE 3.14 Sellafield clean-up

Source: Sellafield Ltd

on to production and to plead for new plants (MOX the latest example) when the need for them is distant or diminishing.

Sellafield is the largest nuclear site in Europe employing around 10,000 workers. Over the years there have been fluctuations and at its peak the plant employed around 17,000 workers. In the early years of this century there were concerns about imminent decline but as the commitment to clean-up has gained priority backed by government investment, the Sellafield works have the prospect of stability in jobs for the next 30 years or so followed by a slow decline lasting into the next century. Since 2011 the priority given to decommissioning and clean-up at Sellafield 'provides higher employment levels for a longer period of time than would otherwise have been the case' (Sellafield Plan, Socio-economic, 2011, p.2). Even so jobs remain top of the agenda for the Sellafield trade unions. 'We need a commitment to Sellafield in the form of guarantees of jobs and socio-economic development' according to Edwin Dinsdale (trade union representative, interview, 2014). Support for new build at adjacent Moorside is strong in West Cumbria but, 'Why the hell would you build one here?' asks Richard Griffin, Head of Strategic Nuclear Policy at Cumbria County Council (interview, 2014). The reason, quite simply, is jobs. 'Policy is driven by economic regeneration. This drives views and top of the list is the economy'.

Sellafield receives around £1.7 billion annually from the government representing over half the total budget of the NDA. Around a third, £600 million, goes into remunerating the Sellafield workforce and about £800 million into the supply chain, with less than 30% of this retained in the West Cumbrian economy (Sellafield Plan, 2011). There has been substantial criticism both of the way this huge investment has been managed and distributed. Following a report of the National Audit Office in 2012 (NAO, 2012) raising some concerns about whether Sellafield was delivering value for money, the management and performance of the Sellafield site received withering criticism from the Public Accounts Committee in 2013 (PAC, 2013). The Committee found 'an extraordinary accumulation of hazardous waste, much of it stored in outdated nuclear facilities'. Criticism of management was especially directed at the fragmentation that had occurred after the collapse of BNFL. In particular the PAC questioned the NDA's appointment of a parent body organisation (PBO), Nuclear Management Partnership (NMP), and a consortium of private sector companies, mostly from overseas, to oversee and improve the performance of Sellafield Ltd, the site operator. Despite cost escalations, poor leadership, poor performance, failure to demonstrate value for money and inadequate protection of taxpayers' interests, the NDA had extended the contract of NMP, a 'highly questionable' decision according to the PAC. Overall there had been little improvement at Sellafield (PAC, 2014). The NMP's £9 billion contract was subsequently terminated. A rather chastened Sellafield Ltd. produced a plan which 'reflects our improved understanding of our task on site. We are more informed about what is deliverable' (Sellafield Ltd., 2013–14, p.11). Certainly, the somewhat simplistic external image of Sellafield as a hazardous facility, poorly managed and devouring large amounts of taxpayers' money was not improved by these reports. Nevertheless, Sellafield now claims to have a clearer focus and the prospect of long-term stability. Adrian Simper summed it up thus: 'There is a hundred years of going forward. A commitment to clean up and an important mission to carry out. There is no future in reprocessing. Employment is stable and the new priority is clean up' (interview, 2014).

The distributive impact of Sellafield on the West Cumbrian economy is a matter of debate. Apart from income generated by the Sellafield workforce and the local supply chain there has been an annual investment of £10 million in projects in the local economy derived from the industry. One of Sellafield's strategic objectives is, 'To support the creation of dynamic, sustainable local economies for communities living near our sites' (Sellafield Strategy, undated, p.26). Over the years this has resulted in support for projects such as Westlakes Science and Technology Park, the Albion office development and the Beacon museum both in Whitehaven, as well as working with small and medium sized enterprises and other local initiatives. Most of this community investment goes into projects which are related to Sellafield itself. Westlakes houses the NDA and university campuses (Central Lancashire and Manchester) specialising in nuclear related activities; the Albion office complex accommodates about 1,000 Sellafield staff transferred from the site; and the museum hosts 'The Sellafield Story', an exhibition on the history of Sellafield which was developed after the closure of the visitor centre on the Sellafield site, at the time one of the most visited tourist attractions in Cumbria. But, the nuclear industry's commitment to the local community is expressed in other ways too, for example in its provision of £500,000 and practical help to the county in the aftermath of flooding on 11 December 2015.

Despite this, there is a sense that Sellafield, especially since the demise of BNFL, has become a more isolated and self-interested benefactor. Tim Knowles, formerly Head of Corporate Affairs with BNFL, claimed, 'In the past it worked actively with local government and the community. It worked like a charm' (interview, 2004). Since then, the fragmentation caused by contractorisation and by the tripartite management of NDA, NMP and Sellafield Ltd is perceived to have diminished the commitment to the community. 'There is no strategic context and companies have no commitment. BNFL had community ethos in the old days.' Although there is a touch of the roseate hues of nostalgia here, the feeling of uneven economic development is common among the leaders of West Cumbria I interviewed in 2014. This was admitted by Adrian Simper, NDA's Director of Strategy. 'It is a quite distorted economy. Sellafield has £1.7 billion but little sticks in Copeland.' Eddie Martin, former Leader of the County Council, emphasised the resulting inequality, 'Copeland has some of the highest wages in Cumbria but Sellafield sits beside people living in abject poverty. We've borne the burden and are paying the penalty for living next door to Sellafield. We have a prehistoric infrastructure.' And his successor as Leader, Stewart Young, reinforced the point. 'We have some of the most deprived communities living in the shadow of a multi-million pound industry. It's an absolute scandal' (interviews, 2014).

Others point to the poor social infrastructure, the low educational standards and condition of the schools, poor health, the threat to hospital services and the lack of cultural facilities. Various studies have identified absence of well-being and multiple deprivation caused by economic insecurity, unemployment or low pay affecting certain parts of West Cumbria (for example, Blackman and Jennings-Peel, 2007; Cumbria Strategic Partnership, undated; ERM Economics, 2003). Community leaders recognise the paradox of poverty in the shadow of a nuclear leviathan. There is resignation and resentment at the lack of resources available to local government to support the social infrastructure. When I last visited in 2014, Copeland, which supplies nearly 60% of Sellafield's workforce, had begun to cut back on even basic services like public toilets. Elaine Woodburn, Leader of the Council, commented, 'There's a £2 billion facility down the road and we can't even cut our grass or keep our centres open'. And there is resentment that councils do not enjoy the full benefit of the business rates that are levied on companies, notably Sellafield (though a reform of business rates including allowing local authorities to retain a proportion of what they collect was put in train in 2015). Pat Graham, Director of Economic Growth at Copeland, summed up the contradictions in the West Cumbrian economy:

> Sellafield is the economic driver for the community, paying £600 million in wages and with 10,000 employees plus contractors. You cannot separate Sellafield from our economy. There is an issue of haves and have nots, those in and those out. The gap is widening. Copeland has the highest wages and cheapest houses with pockets of deprivation, some of the worst health and disease in the UK, an ageing workforce. There is a mass of contradictions with extreme wealth and extreme poverty.
>
> *(Interview, 2014)*

In truth, West Cumbria, or at least the Copeland part, is economically monocultural. Although agriculture, fishing, small manufacturers and public services are all necessarily present, the nuclear industry's presence is pervasive. Over a third of the country's nuclear workforce is in West Cumbria and the combined direct and indirect workforce of Sellafield accounts for 22% of the total in West Cumbria and nearly half (47%) of Copeland's. This dominance and dependence is reinforced by the effective barrier that Sellafield effectively erects against inward investment. Companies from outside the area are deterred by Sellafield's high wages and monopoly of available skilled labour. Thus the two features of the periphery – geography and economy – conspire to sustain Sellafield's hegemony within West Cumbria. Jill Perry of the Green Party and Friends of the Earth describes the effect: 'Sellafield skews the economy, unbalances it. It's difficult to separate the two factors, one the Sellafield effect, two the remote location'.

Not surprisingly a more diversified process of economic regeneration has been hard to achieve. With its declining industries, West Cumbria has been included in the various regional development and assisted area designations over the years and recipient of European Union structural and investment funds. It has also been part of various partnerships, for example the West Cumbria Partnership within the

Cumbria Strategic Partnership and latterly the creation of the Energy Coast Innovation Zone, a public/private partnership comprised of the local councils and the nuclear industry set up 'to build on the area's unique nuclear expertise and global position to develop a dynamic wealth-creating economic future for today' (Britain's Energy Coast Cumbria, 2012). Despite efforts at diversification and a more balanced economy, the inescapable fact of West Cumbria is Sellafield's implanted economic dominance. It is a paradox of power. On the one hand, Sellafield drives a monocultural and unevenly developed economy that is relatively weak in its ability to diversify and draw inward investment. On the other, the very dominance of Sellafield and its importance to the nation provides West Cumbria with the leverage to ensure economic stability. The relationship of dominance and dependence between the nuclear industry and West Cumbria works both ways; West Cumbria depends on the nation and the nation likewise depends on West Cumbria.

Social resignation, resilience and realism

West Cumbria's geographical isolation and economic dependency are important factors in the constitution of the social relations between Sellafield and the West Cumbrian community. The relationship, as Wynne et al. found in their detailed study of *Public Perceptions and the Nuclear Industry in West Cumbria* (1993), is complex: really a set of overlapping, interdependent, interrelated and multivalent attitudes, beliefs and understandings difficult to disentangle. Nevertheless, it is possible to identify different strands of feelings that, taken together, help to explain the sometimes contrasting, contradictory or inconsistent attitudes to the nuclear industry within West Cumbria. As Jean McSorley (interview, 2014) says of personal views of Sellafield: 'It is not a uniform view – within each person there is a massive contradiction'. Despite the complexity, it is possible to detect an underlying realism about the ambivalent relationship between West Cumbrian society and the Sellafield nuclear complex. These various feelings and attitudes, positive and negative, committed and indifferent, may be considered under two contrasting but related concepts identified by Wynne et al., namely, resignation and resilience.

The notion of resignation comprises a number of related feelings, some passive such as fatalism, pessimism or indifference, some more active such as resentment, anxiety or distrust. This leads on to a belief that there is an inevitability about West Cumbria being dumped on. This point was made to me by two Labour leaders of the county council speaking a decade apart. Rex Toft back in 2004 commented: 'We see ourselves as victims of a level of inevitability here – to give us a repository we should be grateful for'. A decade later Stewart Young made a similar point on the subject of a GDF, 'One reason they're putting it there is an acquiescent community' (2014). Critics attribute this feeling of subordination to Sellafield's economic dominance. Martin Forwood of CORE, a long time critic of reprocessing at Sellafield, comments: 'The concentration of activity at Sellafield is scary. It is an unhappy place, a major employer with no real alternative. They have caved in and won't say anything, won't question. People accept it, it passes down the generations' (interview, 2014). His critique is echoed by Jean McSorley: 'At Sellafield people have a huge unease. People suppress the fears they have and problems they can have with it. This results in fear and ambiguity. It is a deprived area and people will put up with crap'.

Certainly, feelings of rejection, inferiority and victimhood help to explain the apparent contradiction between almost universal acceptance of Sellafield and its long-term mission of reprocessing followed by clean-up and the rather more equivocal attitudes taken at the prospect of a deep disposal project in West Cumbria. The sense of being rejected translated, in its turn, into a reciprocal feeling of rejecting Nirex and its putative RCF at the end of the last century. Nirex was seen as an outsider, an unwelcome intruder almost casually and arrogantly disregarding the need for acceptance and integration into the nuclear embrace of West Cumbria. 'Whereas BNFL was recognised as a long-established local company, Nirex was seen as an outsider that had only finished up in West Cumbria because it had failed to find acceptance anywhere else' (Wynne et al., 1993, p.39). On the issue of the GDF, community opinion was fragmented. Those opposing the move to the next stage in the siting process expressed feelings of mistrust, resentment and resignation summed up by Dave Siddall, a local journalist who had reported on Sellafield for many years:

> On the GDF, anyone with any nous knows it's going to be here. The NDA still own the land, 90% of the waste is here. If it went anywhere else there's the logistical nightmare of shifting it. It would not be resisted locally – the councils would roll over and take the bribes. But, the groundswell feels as if it's going to happen anyway. They know there'll be bribes from the nuclear gravy train.
>
> *(Interview, 2014)*

There is another apparent contradiction in the rejection of Nirex and, to a much lesser extent, of the GDF process, and the apparent acceptance of imported spent fuel and the resulting plutonium and vitrified wastes of foreign origin currently stored at Sellafield. An obvious reason for the community's acceptance of foreign nuclear materials at Sellafield is that it was an integral part of the rationale for THORP and the jobs and employment welcomed by the community. By comparison with the repository project for the permanent disposal of radioactive wastes in West Cumbria these imported foreign nuclear materials were originally intended to be resident in the UK for a short period after reprocessing before repatriation as plutonium or in the form of radiologically equivalent high level wastes substituted for the much higher volumes of wastes arising. However, it has subsequently become obvious that repatriation will take a long time and, for some of the foreign-owned stock, may never take place. While repatriation of the wastes has become rhetorically sacrosanct, speaking as it does to ethical principles of self-sufficiency and responsibility, it raises other issues, notably the risks that are posed to the environment and countries en route from the transport of dangerous radioactive materials. It may be argued that these risks outweigh those arising in the UK from continuing to manage the materials which add a relatively small amount to the already large stockpiles of plutonium and wastes. In any case the policy has already been breached through some small transfers of title which raises the question of whether it should be fundamentally reviewed for both pragmatic and ethical reasons. Such a shift might be contested but will be approved in some quarters. Martin Forwood of CORE has long been an advocate of retaining foreign plutonium and waste in the UK:

> Our position is that waste must be kept where it is. No more imports. What we've got, keep and draw a line under it all. The idea of substitution was to minimise transport. Why not follow it to its logical conclusion and keep it? It makes sense to draw a line – stop reprocessing, don't send the waste back, deal with what has to be dealt with.
>
> *(Interview, 2005)*

This kind of attitude reflects the more resilient side of West Cumbrian social response to Sellafield. Resilience brings together notions of pragmatism, optimism, pride and flexibility. Adrian Simper talked to me of Sellafield's responsiveness 'at every call' to its changing fortunes and outlook and the difficult process of adjustment to its clean-up function. There was also the rather stoical attitude of 'We're stuck with the place so we must make the best of it' recorded in Wynne et al.'s study (1993, p.35). The trade unions gave the most forthright expression of this sense of reluctant commitment applied to the prospect of a repository. 'We have a responsibility. We have to knock at the door of No. 10 and say: "We'll take the monkey off your back and this is what we want in return"' (interview, 2005). There is, on the one hand, a sense of obligation and responsibility and, on the other, a belief in entitlement to compensation. Thus, Martin Forwood's (2006) claim, 'Wastes in West Cumbria are our responsibility' is complemented by the trades union comment that 'It's the nation's waste – a service this area has done for the nation'. Consequently, 'If we are going to keep it on site then we need compensation' (Peter Kane, Labour Councillor, Copeland, 2014). This view was strongly expressed to me on several occasions by a range of people.

Both qualities, resignation and resilience, help to account for the various attitudes to the nuclear industry in West Cumbria. The near universal support for Sellafield itself emanates from an awareness of the community's inevitable and umbilical dependence on the nuclear industry refined by a rather defensive assertiveness of West Cumbria's commitment to a project of national importance. This combination of attitudes accounts for the ironic self-awareness so frequently encountered. The relationship between plant and people has semi-feudal overtones. 'They have their fingers in every pie' (Elaine Woodburn, 2014). 'Without daddy/squire we are in a free market' (John Baker, 2005). 'West Cumbria has been dominated by the industry so long that it's entrenched in the fabric' (Martin Forwood, 2014). 'It seeps into the pores of society' (Dave Siddall, 2014). Withal such comments indicate a sense of realism summed up by Wynne et al. (1993, p.59):

> Realism about uncertainties, about lack of power and control, and about dependency, was mitigated by positive recognition of the industry's vital role in the area, and by strong social ties in a variety of different networks of support and identity in the West Cumbrian area.

The reality of power

The interdependent characteristics of geographical isolation, economic dependency and social ambivalence I have just analysed provide the context for the power relations encountered in West Cumbria. Inequalities of power coexist at different levels and the relations are interdependent, encompassing relations between Sellafield and West Cumbria and the wider world. Thus, at one level, within West Cumbria Sellafield appears to be almost omnipotent while on the broader stage its fate appears outside its own control. Similarly, the political and community leadership of West Cumbria tends to project a consciousness of powerlessness with respect to Sellafield on to its relations with the rest of the country. And the two perspectives of Sellafield and of the community of West Cumbria feed into each other; both contrive to portray their relations in terms of power inequalities, controlled and determined by others. In West Cumbria's case, as Wynne et al. observe: 'This sense of a lack of power to influence may have issued in a greater feeling of dependency and fatalism about the nuclear industry and its development' (1993, p.43).

There is a general perception, presumption even, that Sellafield exercises a controlling and dominant power in the region. There is the view that given its predominance in employment in Copeland, its involvement in community organisations, investments and politics, Sellafield holds West Cumbria as some kind of fiefdom. Martin Forwood (2014) put this notion bluntly: 'Sellafield is a mafia – it controls everything, it controls politics, schools, the legal system, cubs and scouts. It is long established and big'. This idea that Sellafield is powerful by its mere presence has echoes of Matthew Crenson's (1971) analysis of United States Steel in Gary, Indiana, USA or my own of the London Brick Company in Bedfordshire, UK (Blowers, 1984). Both were companies whose reputation for power within their communities was such that they appeared to have no need to take any action to make it effective and were impregnable to challenge. This thesis of non-decision making only convinces up to a point. Issues may lie dormant, perhaps suppressed or unrecognised, but eventually they tend to surface and become politicised requiring conscious exercise of power for their determination. There have been many instances where Sellafield has been challenged and forced to act, the events of 1983 being an obvious example. But, the idea of passivity and non-decision making does help to explain the perception and expectation of routine and unchallenged domination that Sellafield appears to hold within West Cumbria.

When it comes to the wider world, Sellafield does not appear to hold uncontested influence. Adrian Simper claims that 'Sellafield is big on the international stage. It transcends boundaries with a big geopolitical footprint in global realpolitik' (interview, 2014). Certainly Sellafield's sheer size as a nuclear complex makes it necessarily a dominant player in the world of nuclear politics involved in transfers of expertise, materials and knowledge alongside other major companies especially those in France and the United States. But, Sellafield is only part, albeit the predominant part, of a state financed organisation, the NDA, engaged mainly in clean-up and waste management operations. It has a client relationship at arm's length to government and, as we have seen, has been open to challenge, criticism and reorganisation. Sellafield, which appears all powerful in its own West Cumbrian bailiwick, seems distinctly inferior when it comes to its relationship with government. Crucial decisions affecting Sellafield's operations and workforce are taken in London. Harold Bolter remarked how Sellafield's management 'see themselves as separate and unloved, ignored when things are going well and bitterly attacked by their colleagues and the media when things go wrong' (1996, p.45). This London decision centredness was remarked on by Jean McSorley (interview, 2014), 'The real decisions are made at Westminster'.

The view of a supplicant community treated with disdain by remote decision makers is widely shared. 'The more compliant and desperate we are, we'll get treated like beggars. The best deal is not going down to London on our knees' (interview with Cumbria CC Nuclear Issues Group, 2006). This is coupled with the view that West Cumbria is treated with disdain by incompetent and uninterested government departments. For instance, Tim Knowles, a Cumbria County Councillor representing a West Cumbria seat, complained:

> The responsible government department, DECC, is weak. Whitehall is incapable of thinking strategically. When you put up an authority like Copeland against government departments backed by big companies then distrust is rife. There has been a failure of government to sustain its policies, to recognise the needs of its local communities. They have serial incompetence, and don't have capabilities of delivery.
>
> *(Interview, 2006)*

The portrayal of inequalities in power relations within West Cumbria is only part of the story. It is taken by many commentators as axiomatic that Sellafield itself assumes a powerful position in the community but that its operations and resources are controlled from outside. Likewise the West Cumbrian community is, in economic and so to an extent political terms, in a client-like relationship to Sellafield while, at the same time, feeling neglected and constrained by Whitehall which holds the purse strings of local government expenditure. Certainly the uneven development I described earlier may be ascribed to these unequal power relations.

Yet, neither Sellafield nor West Cumbria are powerless. The evidence suggests a more nuanced set of relations. After all, throughout its history Sellafield has been the focus of the UK's nuclear industry, the site which, above all others, has claimed the greatest share of investment and the most diverse range of operations. It has been a centre for experiment and innovation and also of multibillion pound investment in projects such as THORP and MOX which failed to meet their overblown expectations. And, as its mission has shifted to decommissioning and clean-up, Sellafield continues to receive government support and subsidy which, though criticised and questioned from time to time, is pretty well guaranteed for years to come. The fact is that Sellafield has enormous leverage derived from its performance of an essential task, cleaning up the nation's nuclear legacy. For West Cumbria, as we have seen, this economic preeminence works both ways. On the one hand, it brings jobs and investment into the area; on the other, it creates inequalities revealed in impoverishment and sustained by disempowerment. From Copeland Borough's perspective, its political impotence in relationship to Sellafield is obvious: 'We don't have leverage. The more we have the more they take advantage' (Elaine Woodburn). 'If we didn't have Sellafield then they (government) would wash their hands completely' (David Moore, Opposition Leader).

Organisations within civil society are somewhat outside the Sellafield nexus. From time to time, they have proved remarkably successful in mounting focused and effective campaigns. Undoubtedly the assault by Greenpeace in Sellafield in 1983 marked a turning point, shining an unwelcome spotlight on reprocessing. And the locally based relentless campaign of CORE exposing failings and problems of reprocessing has helped to raise awareness. But, the overall failure of reprocessing has been brought about by economic and technical failings of Sellafield's own making. NGOs contributed to the downfall of Nirex in 1997 and were instrumental in building up the campaign that ultimately led to the defeat of the government's plans for a GDF in 2013. But, in an area so avowedly pro-nuclear and dependent as West Cumbria, NGOs, with the notable exception of Radiation Free Lakeland, have neither sought nor been able to disturb the mutually supportive relationship between West Cumbria and Sellafield on the issue of new nuclear power stations at Moorside.

At the regional level, West Cumbria's political solidarity with Sellafield has brought it into conflict with the surrounding areas of Cumbria county. While both Copeland and Allerdale Borough Councils were solid in their support of proceeding with the GDF process, Cumbria County Council and the CALC (Cumbria Association of Local Councils) were against. While West Cumbrian councils regard the nuclear industry as an asset, areas further away, in the Lake District, Solway Plain, the Eden Valley and Carlisle, have far less commitment. At the time of the decision on the GDF, the county of Cumbria was administered by two tiers of local government. The county council's veto on proceeding to the next stage was decisive despite the support for the proposal of the two West Cumbrian district councils. Cumbria, as Stewart Young pointed out, 'is supportive of the nuclear mission, decommissioning and new build. We're not anti-nuclear'. But, the county council was sceptical about the GDF and gained little assurance from the government on such issues as the right to withdraw from the process or on community benefits. Likewise CALC, representing the lowest tier of government, the parish councils, had also agreed not to move forward on the grounds that certain conditions had to be met.

The division of powers within Cumbria had frustrated the government's hopes for eventually finding a site in the county for the GDF, a point that had not gone unnoticed. In its subsequent consultation on the siting process, the government disingenuously indicated that, in future, the final decision on whether to proceed should be at the level of the district council in England with the county council able to influence as a member of a 'Consultative Partnership' (DECC, 2013). To many respondents this smacked very much of re-entering West Cumbria by the back door with the front gate firmly shut to the county. Eventually, the government backed off and in a White Paper declared that 'no one tier should be able to prevent the participation of other members of that community' (DECC, 2014, p.43) leaving the problem of who should be able to volunteer in any future siting process to a Community Representation Working Group of which, incidentally, I became a member.

Living with risk

Sellafield is synonymous with risk – radioactive, financial and technological. It conjures up in the popular imagination a dangerous place where the risks from radioactivity to humans and the environment are a quotidian reality. The impact of radioactivity on people's health has been a continuing debate over evidence notably recorded in the Black Report (1984), in the media and in the personal accounts recorded by Sally Macgill (1987), Marilynne Robinson (1989), Jean McSorley (1990), Brian Wynne and colleagues (1993) and Hunter Davies (2012). Sellafield's image, developed through investigations and the revelations of hostile NGOs and conveyed through the national media, is of an organisation poorly managed, with an insatiable appetite for public funds, its finances out of control. It appears to have a perverted genius for developing technology that fails to reach planned expectations. Over the years Sellafield has been prone to radioactive releases, some serious, and a reputation for secrecy and cover up. It may be said that the evidence gives support to the unfavourable image. The Windscale fire of 1957, the leaking tanks of 1976 kept secret until 1979, the discharges into the Irish Sea and beach in 1983, the woeful underperformance of the MOX plant (producing five tonnes in its first five years with an annual capacity of 120 tonnes), the failure of THORP to achieve even the most modest throughputs are but the most prominent of a litany of leaks, cover ups, major incidents and failures. The scathing Parliamentary criticisms of financial and managerial weakness and the changes in leadership and organisation betoken an operation that seems almost out of control. Over the years Sellafield has unwittingly cultivated and endured a negative image. The hazardous legacy wastes were described by Margaret Hodge, Chair of the Public Accounts Committee, as posing 'intolerable risks to people and the environment' (BBC News, 7 November 2012).

Once implanted, an image may be difficult to dislodge, especially if it is provided with constant fulfilment. The more positive features of Sellafield – its commitment to safety, good labour relations and, above all, its engagement in a difficult and dangerous task that lacks any kind of glamour or recognition for achievement – are largely unregarded in the wider world. But for the workers of Sellafield and for the community at large the risks of working and living there are accepted as obligatory. There is an awareness of risk but an acceptance that it is unavoidable and a sense of responsibility on the part of the local community that the risks have to be safely managed on behalf of the wider community. It is this sense of perceiving the risks and understanding the responsibilities that enables the community to live with Sellafield.

Sellafield's relationship with its local community is ambivalent. Its economic importance and power are recognised and there is a degree of loyalty and pride especially among the workforce in Sellafield's mission though less so as its mission has shifted from plutonium to clean-up. But, there is another, darker side to the relationship borne of Sellafield's reputation as a place of danger marked by incidents, cover ups and secrecy. This has encouraged suspicion and reserve in its relations with parts of the local community. There is, to be sure, mutual interest but it is sometimes tinged with mutual distrust. The point was put to me vividly by Richard Griffin in 2014,

> I always felt that BNFL hadn't told you everything. The NDA came in on the back of this secretive attitude on the eve of openness and transparency but they behaved much as BNFL had done. The Freedom of Information Act had changed the 'need to know' to 'the right to know' and this should change the way you behave. In local councils the lack of trust is still there right through to the MRWS.

While there is a wariness in relations between Sellafield, whether under BNFL or the NDA, the distrust between West Cumbria and outsiders borders on hostility. Certainly, Nirex endured a desperate reputation, making almost no effort to nurture community relations. By contrast, considerable effort was made by government with the MRWS process pioneered by CoRWM to develop a partnership approach, raise community awareness and encourage engagement. Even so there was ultimately a failure to inculcate a basis of trust, a key reason why Eddie Martin led the county council to reject going forward with the GDF process. 'We don't trust them. The kind of things we've heard don't inspire us. The level of trust is at a significantly low ebb. I think they'll find an excuse for coming back to Cumbria' (interview, 2014).

At least, it may be said, Sellafield is deeply embedded in the community, not a carpetbagger from outside. Efforts through its socio-economic programme, community involvement and public relations help to promote what Sally Macgill has called a 'politics of reassurance'. But while these efforts may diminish the 'politics of anxiety' that for long pervaded Sellafield, it cannot totally remove the

FIGURE 3.15 Sellafield cartoons

Source: Martyn Turner/Irish Times

stigma which naturally attaches to a community which has, at its heart, a notorious nuclear leviathan. Living with Sellafield requires adaptation to the tendency to self-denigration it implies. The local community seems to have achieved this through a combination of denial and self-deprecation. This rather subtle, even ironic, approach to the outside world was, for me, brilliantly summed up by an epigrammatic lapel badge I was given long ago and which I still proudly own. It says, simply, 'I've been to Sellafield!'.

Visits and interviews

I first became involved in radioactive waste management in October 1983 when I was a County Councillor in Bedfordshire. and Elstow in that county. was identified as a possible site for a nuclear waste repository. I became involved in the campaign led by the council and the local community to defeat the proposal. Subsequently, I was appointed to the government's Radioactive Waste Management Advisory Committee (RWMAC) between 1991–2003 and then the first Committee on Radioactive Waste Management (2003–07) which recommended the policies for radioactive waste management discussed in this chapter. I was also, for a time (2000–04), a non-Executive Director of UK Nirex Ltd., the company responsible for managing the wastes until it was absorbed into the Nuclear Decommissioning Authority (NDA). From 2008 I set up an anti-nuclear campaign group, Blackwater Against New Nuclear Group (BANNG) and later became Co-Chair of the Department of Energy and Climate Change/NGO Nuclear Forum. In all these roles I was involved in the development of policy and met many people in the industry, in government, regulators, politicians, academics, environmental groups and citizens who, though nameless, have all contributed to this chapter. I thank them all.

Over the years I have made several visits to Sellafield and West Cumbria. During some of these visits I interviewed people who are named below and many of whom are quoted in this chapter.

1993–94 interviews as part of Global Environmental Change Open University Project conducted in London, Kendal, Sellafield, Barrow-in-Furness and Milton Keynes

Interviewed: Peter Wilkinson, former Greenpeace; John Hetherington and Windsor Briggs, Cumbria County Council; Bob Phillips, Sue Larkins, John Barbour and Duncan Jackson, BNFL; Alan Westonridge, trade unionist; Martin Forwood and Janine Allis-Smith, CORE; Patrick Green, Friends of the Earth; Susan Meyer, Greenpeace.

2004, April, Sellafield and West Cumbria, CoRWM

Interviewed: John Clarke, Environmental Health, Quality, Safety, Sellafield; Rex Toft, Leader and Tim Knowles, Councillor, Cumbria County Council; Councillors Geoffrey Blackwell and Elaine Woodburn, Leader Copeland Borough Council; Martin Forwood, CORE.

2005, January, Sellafield, West Cumbria, CoRWM

Interviewed: Norman Williams, Geoff Blackwell, Councillors and portfolio holders; Cllr. Norman Clarkson, Deputy Mayor, Copeland Borough; Fergus McMorrow, Director, Economic Prosperity and Sustainability, Copeland Borough; John Grainger, Cumbrian Inward Investment Agency; Rina Barber, West Cumbria Strategic Partnership; John Baker, Whitehaven Community Trust.

2005, November, December, West Cumbria, CoRWM

Martin Forwood and Janine Allis-Smith, CORE; Anita Stizecker, South Lakes Friends of the Earth; Marjorie Higham (Drigg); Marilyn Tahernia (FoE), trade union representative; Keith Bradshaw, West Cumbria Partnership; John Hetherington, Fergus McMorrow and Sue Stephenson, Cumbria Strategic Partnership.

2006, May, West Cumbria, CoRWM

David Davies and Sue Stephenson, Cumbria Strategic Partnership; Marilyn Tahernia and Anita Stizecker, FoE; Marjorie Higham (Drigg); Martin Forwood, CORE; Howard Rooms (UCATT), Peter Kane, (GMB), Paul Shawcross and Liz Lamb (Prospect), Sellafield Trade Unions.

2014, October, Barrow-in-Furness, West Cumbria, Sellafield and Carlisle – research visit

Interviewed: Jean McSorley, Barrow-in-Furness, former Greenpeace; Martin Forwood and Janine Allis-Smith, CORE; Richard Griffin, Strategic Nuclear Policy, Cumbria County Council; Dave Siddall, freelance journalist; Eddie Martin, former Leader Cumbria County Council and Cumbria Trust; Geoff Betsworth, Colin Wales and Rod Donington-Smith, Cumbria Trust; Jill Perry, Friends of the Earth; Cllr. Elaine Woodburn, Leader and Councillors Peter Kane, Alan Holliday, David Moore, Copeland Borough Council; Pat Graham, Director Economic Growth, and John Groves, Nuclear and Planning Manager, Copeland Borough; Tim Knowles, County Councillor and former Head of Corporate Affairs, BNFL; Chris Shaw and Keith Hitchen, Cumbria Association of Local Councils; Brian Hough, Stakeholder and Socio-economics Manager, NDA, Cumbria; Adrian Simper, Director, Strategy and Technology, NDA; Edwin Dinsdale, GMB Union Sellafield; Cllr. Stewart Young, Leader, Cumbria County Council.

Notes

1 The stockpile of radioactive wastes comprising the legacy that must be managed in the UK and which is the subject of this chapter comprises several components. High level waste is very radioactive and generates a great deal of heat and accounts for about half the total radioactivity in the UK's inventory. Spent fuel, though not declared yet as a waste, has to be managed and accounts for a further 42% of the radioactivity. Other components in the legacy stockpile are plutonium (5%), intermediate level waste (around 3% activity but around three-quarters of the volume) and uranium low in activity but 15% of the volume (CoRWM, 2006, p.25). Low level waste which comprises by far the largest volume is not included here since it is permanently disposed of in a repository.
2 Sellafield takes its name from a small community where the works were built. Its name means field of willows. In order to distinguish it from Springfields, the nuclear fuel production centre further south in Lancashire, the name Windscale was initially adopted named after a bluff overlooking the river Calder which gives its name to the Calder Hall power station.
3 The division between short-lived and long-lived intermediate level wastes is based on the half-lives of radionuclides of more or less than 30 years.

References

Atherton, E. (2000) *Veto and Volunteerism*, Harwell, United Kingdom Nirex Ltd. UKNL (00) 17.

Aubrey, C. (1993) *THORP: The Whitehall Nightmare*, Oxford, Jon Carpenter.

Bell, C. and Newby, H. (1971) *Community Studies*, London, George Allen and Unwin.

Black Report, The (1984) *Investigation of the Possible Increased Incidence of Cancer in West Cumbria*, report of the independent Advisory Group, chairman Sir Douglas Black, Department of Health and Social Security, London, HMSO.

Blackman, T. and Jennings-Peel, H. (2007) *A Health Impact Assessment of Housing, Worklessness, Children's Services and Primary Care Services*, The Whitehaven and

Workington Neighbourhood Management Initiative Areas, School of Applied Social Sciences, Durham University, July.

Blowers, A. (1984) *Something in the Air: Corporate Power and the Environment*, London, Harper and Row.

Blowers, A., Lowry, D. and Solomon, B. (1991) *The International Politics of Nuclear Waste,* London, Macmillan.

Blowers, A. and Lowry, D. (1993) 'THORP: to be or not to be', *Safe Energy,* 95, June/July, 16–17.

Blowers, A. and Lowry, D. (1996) 'Nuclear transportation – the global vision', *Proceedings of International Topical Meeting on Nuclear and Hazardous Waste Management*, Seattle, Washington, August, American Nuclear Society Inc., La Grange Park, Illinois 60526.

Bolter, H. (1996) *Inside Sellafield,* London, Quartet Books.

Breach, I. (1978) *Windscale Fallout*, Harmondsworth, Penguin Books.

Britain's Energy Coast Cumbria (2012) *The West Cumbria Economic Blueprint*, www.britainsenergycoast.co.uk (accessed 7 June 2012).

British Geological Survey (BGS) (2010) *Managing Radioactive Waste Safely: Initial Geological Unsuitability Screening of West Cumbria*, CR/10/072, BGS, Nottingham.

Brown, K. (2013) *Plutopia: Nuclear Families, Atomic Cities, and the Great Soviet and American Plutonium Disasters*, Oxford, Oxford University Press.

CALC (Cumbria Association of Local Councils) (2012) *Managing Radioactive Waste Safely*, Views of the Cumbria Association of Local Councils, August.

CORE (undated) *THORP*, Barrow-in-Furness, Cumbria.

CoRWM (2006) *Managing our Radioactive Wastes Safely: CoRWM's Recommendations to Government*, London, CoRWM, November.

CoRWM (2007a) *Ethics and Decision Making for Radioactive Waste*, A Report for CoRWM, edited by Andrew Blowers, February.

CoRWM (2007b) *Moving Forward: CoRWM's Proposals for Implementation*, CoRWM document 1703, February.

CoRWM (2007c) *Implementing a Partnership Approach to Radioactive Waste Management – Report to Governments*, CoRWM Paper 2146, April.

Crenson, M. (1971) *The Un-Politics of Air Pollution: A Study of Non-Decisionmaking in the Cities*, Baltimore, The Johns Hopkins Press.

Cumbria County Council (1996) *Appeal by United Kingdom Nirex Ltd*, Inspector's Report, 21 November.

Cumbria Strategic Partnership (undated) *Sustainable Cumbria 2004 to 2024. A Strategy for Growth and Progress for Cumbria.* Prepared for the CSP by John Glester Consultancy Services.

Davies, H. (ed.) (2012) *Sellafield Stories: Life with Britain's First Nuclear Plant*, London, Constable.

DECC (2011) *Management of the UK's Plutonium Stocks, A Consultation Response on the Long-term Management of UK-owned Separated Civil Plutonium*, London, December.

DECC (2013) *Review of the Siting Process for a Geological Disposal Facility*, Consultation, DECC, September.

DECC (2014) *Implementing Geological Disposal, A Framework for the Long-term Management of Higher Activity Radioactive Waste,* White Paper, DECC, July.

DECC (2015a) *Foreign Plutonium and Waste – Issues of Transfer, Substitution and Storage – Questions for Consideration: Spent Nuclear Fuel and Reprocessing in the UK*, papers presented at the DECC/NGO Nuclear Forum, January.

DECC (2015b) *Plutonium Management in the UK – Title Transfers*, paper presented to the DECC/NGO Nuclear Forum, July.

Defra (2001) *Managing Radioactive Waste Safely – Proposals for Developing a Policy for Managing Radioactive Waste in the UK,* Department for Environment, Food and Rural Affairs, Defra, September.

Defra (2006) *Response to the Report and Recommendations from the Committee on Radioactive Waste Management, UK Government and the Devolved Administrations,* Defra, October.

Defra (2007) *Managing Radioactive Waste Safely: A Framework for Implementing Geological Disposal,* a public consultation by Defra, DTI and the Welsh and Northern Ireland Devolved Administrations, London, Defra, June.

Defra (2008a) *Managing Radioactive Waste Safely: A Framework for Implementing Geological Disposal,* White Paper by Defra, BERR and the Devolved Administrations for Wales and Northern Ireland, London, TSO, CM 7386.

Defra (2008b) *Summary and Analysis of Responses to the Consultation on Managing Radioactive Waste Safely: A Framework for Implementing Geological Disposal,* Defra, BERR and the Devolved Administrations for Wales and Northern Ireland, January.

ERM Economics (2003) *West Cumbria: A Socio-economic Study – 2003 update,* The Environment Council: BNFL National Stakeholder Dialogue, ERM Economics, June.

Friends of the Earth (1999) *THORP: the Case for Contract Renegotiation,* by Mike Sadnicki, Fred Barker and Gordon MacKerron, London, June.

FWS Consultants (2011) Requested response to Professor Smythe's paper, letter from Dr J. Dearlove, FWS Consultants Ltd. to West Cumbrian MRWS Partnership, 26 October.

Government Office for the North West (1997) *Appeal by United Kingdom Nirex Ltd. Proposed Rock Characterisation Facility on Land at and Adjoining Longlands Farm, Gosforth, Cumbria,* 17 March.

Hall, T. (1986) *Nuclear Politics,* Harmondsworth, Penguin Books.

Haszeldine, S. and Smythe, D. (eds) (1996a) *Radioactive Waste Disposal at Sellafield, UK, Site Selection, Geological and Engineering Problems,* University of Glasgow.

Haszeldine, S. and Smythe, D. (1996b) 'Overview of the site selection, geological and engineering problems facing radioactive waste disposal at Sellafield, UK', in Haszeldine, S. and Smythe, D. (eds), *Radioactive Waste Disposal at Sellafield, UK, Site Selection, Geological and Engineering Problems,* University of Glasgow, pp.1–3.

Herring, H. (2005) *From Energy Dreams to Nuclear Nightmares: Lessons from the Anti-nuclear Power Movement in the 1970s,* Oxford, Jon Carpenter.

Hinchliffe, S. and Blowers, A. (2003) 'Environmental responses: radioactive risks and uncertainty', in Blowers, A. and Hinchliffe, S. (eds) *Environmental Responses,* Milton Keynes, The Open University and Chichester, John Wiley.

House of Lords (1999) *Management of Nuclear Waste,* Select Committee on Science and Technology, Session 1998–99, Third Report, London, HMSO.

Ipsos MORI (2012) *Managing Radioactive Waste Safely Survey 2012,* prepared for West Cumbria MRWS Partnership, May.

Jay, K. (1956) *Calder Hall: The Story of Britain's First Atomic Power Station,* London, Methuen.

Macgill, S. (1987) *The Politics of Anxiety: Selllafield's Cancer-link Controversy,* London, Pion.

Macgill, S. and Phipps, S. (1987) 'The Sellafield controversy: the state of local attitudes', in Blowers, A. and Pepper, D. (eds), *Nuclear Power in Crisis,* London, Croom Helm.

McSorley, J. (1990) *Living in the Shadow: The Story of the People of Sellafield,* London, Pan Books.

Medvedev, Z. (1979) *Nuclear Disaster in the Urals,* London, Angus and Robertson.

Milliken, R. (1986) *No Conceivable Injury: The Story of Britain and Australia's Atomic Cover-up,* Ringwood, Victoria, Penguin Books Australia.

Ministry of Fuel and Power (1955) *A Programme of Nuclear Power,* Cmnd 9389, HMSO.

NAO (National Audit Office) (2012) Managing Risk Reduction at Sellafield: Nuclear Decommissioning Authority, HC 630, Report by the Comptroller and Auditor General, Session 2012-13, November, London, TSO.

NDA (Nuclear Decommissioning Authority) (2014) *Progress on Approaches to the Management of Separated Plutonium*, Position Paper, January.

Nirex (1987) *The Way Forward: A Discussion Document,* Harwell, UK Nirex Ltd.

Nirex (1993) *Scientific Update 1993: Nirex Deep Waste Repository Project*, Harwell.

Nirex (2000) *Transparency Policy*, Harwell, August.

O'Neil, J. (2001) 'Representing people, representing nature, representing the world', *Environment and Planning C (Government and Policy)*, 19, 483–500.

Openshaw, S. (1986) *Nuclear Power: Siting and Safety,* London, Routledge and Kegan Paul.

Phillips, L. (1995) *Multi-attribute Decision Analysis for Recommending Sites to be Investigated for their Suitability as a Repository for Radioactive Wastes*, Proof of Evidence on behalf of UK Nirex Ltd. IN431, October.

Public Accounts Committee (PAC) (2013) *Nuclear Decommissioning Authority: Managing Risk at Sellafield*, 4 February.

Public Accounts Committee (2014) *Forty-Third Report, Progresss at Sellafield*, 3 February.

RWMAC (Radioactive Waste Management Advisory Committee) (1999) *The Establishment of Scientific Consensus on the Interpretation and Significance of the Results of Science Programmes into Radioactive Waste Disposal,* The Radioactive Waste Management Advisory Committee's Advice to Ministers, London, RWMAC, April.

RWMAC (2001) *The Process for Formulation of Future Policy for the Long-term Management of UK Solid Radioactive Waste*, The Radioactive Waste Management Advisory Committee's Advice to Ministers, London, RWMAC, September.

Robinson, M. (1989) *Mother Country: Britain, the Welfare State and Nuclear Pollution,* New York, Ferrar, Strauss and Giroux.

Roche, P. (2015) *Towards a Safer Cumbria*, 2015 Update, Cumbrian Energy Revolution, February.

Sellafield Plan (2011) *Risk and Hazard Reduction, HAL Workstream; Socio-economic*, Issue 1, August, Sellafield Ltd.

Sellafield Ltd (2013–14) *Annual Review 2013–14*, Sellafield Ltd., Seascale, Cumbria.

Sellafield Strategy (undated) *Key to Britain's Energy Future, The Strategy for Sellafield*, www.sellafieldsites.com (accessed 18 April 2016).

Smythe, D. (2011) Response to letter submitted to West Cumbria MRWS by Dr J. Dearlove dated 26 October 2011, 7 December.

UEA (University of East Anglia) (1988) *Responses to The Way Forward,* Environmental Risk Assessmeant Unit, Norwich, November.

Walker, W. (2007) The UK and Nuclear Reprocessing: Beating a Retreat, Public Citizen, www.citizen.org (accessed 18 April 2016).

Walsh, I. (2000) *Mobilising Modernity: The Nuclear Moment,* London, Routledge.

West Cumbria MRWS Partnership (2011a) *Geological Disposal of Radioactive Waste in West Cumbria?* Public Consultation Document, November 2011 to March 2012.

West Cumbria MRWS Partnership (2011b) Letter from DECC regarding the Councils' Memorandum of Understanding, 7 November, doc. no. 240.

West Cumbria MRWS Partnership (2012a) *Public and Stakeholder Engagement Round 3 Interim Report,* a report of the West Cumbria MRWS Partnership's formal consultation and opinion survey, Document 288, June.

West Cumbria MRWS Partnership (2012b) *The Final Report of the West Cumbria Managing Radioactive Waste Safely Partnership,* August.

West Cumbria MRWS Partnership (2012c) Memorandum of Understanding between Cumbria County Council, Copeland Borough Council and Allerdale Borough Council, Doc. No. 235, formerly published as draft November 2011.

Williams, W. (1956) *The Sociology of an English Village,* London, Routledge and Kegan Paul.

Wynne, B., Waterton, C. and Grove-White, R. (1993) *Public Perceptions and the Nuclear Industry in West Cumbria,* Centre for the Study of Environmental Change, Lancaster University.

4

LA HAGUE AND BURE, FRANCE

The traditional and the modern

Picture a harsh, wild stretch of nature with splendid horizons, endlessly fascinating in terms of the variety it offers to the eye. The interior of the peninsula is occupied by a great plateau consisting of a series of dome-like moors where gorse and broom, heather and bracken are swept by incessant wind.

(Françoise Zonabend, The Nuclear Peninsula, *1993, p.13)*

A Bure-en-Barrois vous êtes dans le sud-ouest de la Marne, dans un *no man's land* de 5 a 7 habitants au km² ... Le village est en récession économique de longue date ... Mais l'implantation de ce site devrait s'accompagner d'aménagements et d'équipements péripheriques, entraîner la venue de milliers de personnes et relancer le développement économique local.

(Ben Cramer and Camille Saïsset, La Descente aux Enfers Nucléaires, *2004)*

Introduction – the core on the periphery

Cap de la Hague is at the tip of the Cotentin Peninsula in the Manche department in Normandy, the point where northwestern continental Europe finally runs into the sea. This remote setting is the location for one of the largest nuclear complexes in the world, with its high level waste stores and two large reprocessing plants which separate uranium and plutonium from spent fuel to complete the French nuclear fuel cycle. Together with the nearby Arsenal at Cherbourg, where French nuclear submarines are constructed, and the two Flamanville nuclear power plants down the coast (one with two reactors opened in 1986, the other under construction since 2007 and still unfinished in 2016) this 'nuclear peninsula' directly provides around a fifth of the jobs in the northern Cotentin and supports the infrastructure and dependent activities for a much wider population. La Hague represents a commitment to nuclear energy and technology that is still at the heart of the French nuclear industry.

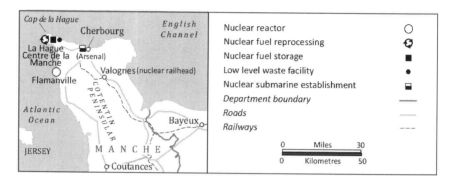

FIGURE 4.1 Map showing the Cotentin Peninsular, northwest France

Map by: John Hunt

FIGURE 4.2 The La Hague nuclear fuel reprocessing plant on the Cotentin Peninsula, Normandy

Source: Truzguiladh WiKi

From La Hague it is 400 miles to Bure in the heart of rural eastern France. Bure, too, is remote, in a thinly populated region, far from major cities and communications. Yet, like La Hague, it is a strategic location for the nuclear industry. For, as La Hague fulfils its function in storing reprocessed vitrified and other long-lived wastes, so Bure is, by degrees and slowly, becoming the place where these wastes may eventually be placed deep underground in a geological repository. As yet the

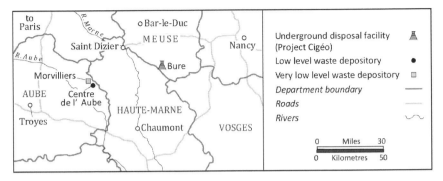

FIGURE 4.3   Map showing the location of Bure and the surrounding region

Map by: John Hunt

FIGURE 4.4   The deep geological disposal facility Cigéo under development at Bure, eastern France

Source: © ANDRA

imprint of the nuclear industry here is confined to some modern structures which appear suddenly and discordantly contrasting with the vernacular farmhouses and stone built villages that are dotted about in the expansive landscape. Yet the destinies of these two relatively obscure places are to be intertwined by their central role in the storage and disposal of France's high level radioactive wastes. Although nothing in the management of radioactive wastes can ever be certain, the inertia that maintains reprocessing at La Hague seems also likely to underlie the continuing commitment to Bure as the place for the final solution of the problem of the more dangerous radioactive waste.

The nuclear industry in France developed a little later than in the US and UK but, it, too, comprises the full panoply of activities, military and civil. In this chapter I shall take a slightly different approach to the previous ones by giving more attention to the broader national policy context that has shaped the development of the two nuclear oases that are the subject of this study. I do so since French nuclear development has, from the outset, been *dirigiste,* an expression of a highly centralised state direction and its concomitant form of somewhat top down elitist decision making. After briefly setting the context of nuclear energy in France in the first part of this chapter, I shall frame my analysis through the lens of nuclear discourses in the second part. In France the nuclear discourses present an overlapping rather than sequential expression with the idea of Trust in Technology firmly established at the outset and subsequently sustained by a powerful combination of business and political relationships. Its dominance was challenged but not displaced, by a discourse of Danger and Distrust embodied in a vigorous anti-nuclear movement able to mobilise mass protest but rarely to achieve palpable success in its overall aim but in some cases able to prevent the siting of radioactive waste facilities. This discursive conflict has been, especially at a local level, moderated by an emerging discourse, a Gallic version of Cooperation and Consensus especially on issues relating to managing the nuclear legacy. After initial efforts to impose sites for radioactive waste disposal on local communities were rebuffed, there has been a detectable shift towards a more open approach, one based on an effort to cooperate with local communities in the search for volunteers. In the French case, as we shall see in the context of the two communities I have chosen, consensus and cooperation has been confined to the representative political, stakeholder and bureaucratic institutions rather than exposed to a more wide-ranging form of public participation.

The shift in both focus and discourse can be traced and understood in terms of the concept of the periphery, the local context of the two communities that are the subjects of parts 3 and 4 of this chapter. La Hague, the focus of the third part, has been closely entwined with development of French nuclear energy and its reprocessing function closes the nuclear cycle and sustains the nuclear project in France. The ability of La Hague to adapt and adjust its role was a key element in the industry's ability for so long to maintain its serene and secure position in the energy sector. La Hague provides a fascinating study of survival and integration within a traditional community that both supports and is supported by the implantation of a modern industrial complex. By contrast, at Bure, covered in the fourth part, the processes of integration are only just beginning. But, like La Hague, Bure has become one of the main bases on which the nuclear industry depends for its continuing justification and survival. Unlike La Hague, Bure is not a nuclear community in the sense of an established population living and working in the area. Rather it is a nascent community which may continue in its present form of scattered settlements but containing a growing adventitious population. Both communities provide an insight into the local implications and impacts of the nuclear legacy on the traditional periphery. In the concluding part of the chapter I shall return to the broader national political and policy context to reflect on the prospects facing the two communities that are increasingly responsible for the management of France's nuclear legacy.

La Hague and Bure are peripheral locations where the presence of the nuclear industry sets in train social processes of modernisation as traditional, one might say backward, communities are transformed and integrated more into the mainstream of modernity. Of course, this is a continuing process with no evident beginning or end since it is so intertwined with other processes, economic, cultural, technological, that together shape and differentiate communities within modern society. But, the introduction of a modern nuclear technology into a previously quiescent community brings a dramatic exposure to the risks and economic and social changes that go with the confrontation and ultimately the integration of the traditional and the modern. These two places are at different ends of a continuum. La Hague has been through the disruptive beginning of the process of change, through a phase of conflict and accommodation to the later stable stages of the process where the modern has become embedded thereby evolving, as it were, into a new traditional, in which the two are at once both contradictory and complementary. While La Hague has completed the journey, at Bure it has scarcely commenced. Bure is at the beginning of the process where the nuclear industry has laid the technical foundations for a permanent presence but has only just begun to touch the social and economic bases that may eventually transform the region. Bure is at a formative stage and it is a matter of speculation as to how far geographically and how deep socially the changes will extend.

In this chapter I will look at the process of peripheralisation as it has impacted on these two communities. La Hague and Bure are the fundamental bases for the end stages of the nuclear cycle – the reprocessing of spent fuel and the long-term management of highly active wastes. And their role needs to be understood within the broader context of French nuclear policy and the interaction of the national and the local through which French nuclear policy has been developed and implemented. I begin by exploring this national context.

Tout electrique, tout nucléaire – the French nuclear cycle

The French nuclear industry is comprehensive, large and purposeful. Its origins lie in the aftermath of the Second World War in the determination to reassert France's autonomy and world role in which nuclear weapons played a part. In the early years there was a focus on plutonium for military purposes located at the reprocessing factory at Marcoule in the Rhône Valley so vividly recorded by Gabrielle Hecht (1998). My concern in this chapter is with the development of the French nuclear energy programme which developed slightly later and more especially with the back end of the nuclear cycle in France, reprocessing at La Hague in Normandy and the consequent implications for clean-up and radioactive waste management. The French nuclear energy programme stems from a single-minded commitment by the state arising from decisions in the 1970s to achieve an independent energy base for France. France has embraced the full spectrum of the nuclear cycle, from the development of nuclear weapons, to the production of nuclear energy, the reprocessing of spent fuel and the long-term management of nuclear wastes. France is second only to the USA in nuclear energy with a generating capacity of 63GW in 2014 accounting for three-quarters of electricity consumption in the country. The scale and comprehensiveness of the French nuclear programme was determined in the aftermath of the Middle East oil crisis of 1973 which exposed the country's vulnerability in energy supply. This led to

> a determination to ensure French control over energy resources and consequently to guarantee their security. The key to the policy lay in a massive expansion of the nuclear programme and the establishment of national control over all aspects of the nuclear fuel cycle.
>
> (Boyle and Robinson, 1987, pp.56–7)

The influence of the nuclear industry in France is pervasive in its geographical reach, technical capability and political power.

Geographically, the country's 58 reactors are located along the Channel coast, the German, Luxembourg and Belgian borders and on the major rivers (Loire, Seine, Rhône, Rhine, Moselle and Garonne) (Davis, 2002). Along the Rhône are the fuel enrichment plants (Pierrelatte and Tricastin). Fuel fabrication is carried out mainly in the southeast and the fuel cycle is completed with the reprocessing at La Hague in Normandy.[1] Some of the plutonium and uranium from reprocessing is used to produce Mixed Oxide Fuel (MOX) at the Melox plant at Marcoule. MOX is used in around half of the country's reactors and provides a continuing justification for reprocessing. The early French flirtation with fast breeder technology as a means of using reprocessed fuels was effectively ended with the closure of the Superphénix plant at Creys-Malville on the Rhône in 1997, though a revival of the technology in a fourth generation of nuclear power stations remains very firmly fixed in the future ambitions of the nuclear establishment.

Inevitably France has a growing burden of nuclear wastes that have to be managed. There has been relatively little controversy over managing the bulky low activity wastes. For 25 years (1969–94) the main disposal site for these wastes was the so-called Centre de la Manche, adjacent to the La Hague reprocessing plant and named after the department in which it was situated (County Councils Coalition, 1987; RWMAC, 1994). This was a surface disposal facility which accommodated 525,000m^3 of short-lived low and medium level waste. Although there were criticisms of the site during its operation, by comparison with the neighbouring reprocessing plant it has attracted little public attention. A similar facility, Centre de l'Aube (also named after the local department), has been operating since 1992 at Soulaines near Troyes in Champagne, eastern France. It was one of three candidates (the others were in Indre and Vienne departments in west-central France) chosen for its suitable geology and low seismic activity.

In the 'closed' French nuclear cycle, spent fuel from nuclear reactors is reprocessed and the high level (HL) waste streams are consolidated into glass blocks. Eventually

FIGURE 4.5   Map of nuclear facilities in France

Map by: John Hunt

the vitrified blocks will be transferred from La Hague for deep underground disposal at a repository with Bure as the most likely site. Such a solution, enshrined in law, appears simple and logical, satisfying the French penchant for clarity and rationality in decision making. However, the actual situation is more complex and uncertain. Bure still has to progress through authorisation and the future inventory and composition of wastes destined for the repository will depend on the future development of the industry. Not all the spent fuel will be required for production of MOX and volumes of unreprocessed spent fuel will, therefore, accumulate at power stations and at La Hague. The future management of used MOX is unclear and there will be stocks of plutonium and uranium surplus to requirements. And, if nuclear energy becomes a declining part of the French energy mix, then the materials presently regarded as potentially usable will have no further use and consequently may become declared as wastes. Thus, the scale and scope of the problem of nuclear waste management in France is indeterminable.

The triumph of technocratic centrism

Nuclear energy and the state

Nuclear energy policy in France is strongly centrist; devised, developed and implemented by the state and its industrial and governmental agencies. For such a vast industry the operating structure is relatively simple. The CEA (since 2009, Commissariat à l'énergie atomique et aux énergies alternatives, to emphasise its renewable energy credentials) is the state's energy research and development organisation for the development of applications of nuclear power and, now, for renewable energy as well. It is the lead body for research into partitioning and transmutation of radioactive waste as a means of reducing the long-lived isotopes it contains. Electricité de France (EDF), a private company with the state taking 85% of the shareholding, operates all the reactors and is, by far, the main producer of nuclear waste. Another company, Areva (formed in 2001 by a merger of COGEMA, the former reprocessing company, Framatome the reactor builder and a part of the CEA)[2] also with the state holding the bulk of the shares, provides a range of fuel services including reactor design, reprocessing, fuel enrichment and fabrication, uranium mining and reactor technology. Areva's financial losses led to the French government in 2015 ordering a merger of its reactor business with EDF in a joint venture controlled by EDF. A third company, ANDRA,[3] founded in 1979, is responsible for managing the country's nuclear wastes under policies determined by law, notably the laws of 1991 and 2006 (see below). France has also maintained a commitment to standardisation of reactor design. After initial investment in gas-cooled reactors, France adopted a version of the Westinghouse PWR system with reactor capacities of 900 MW and later 1,200 MW, with the latest generation heralded by the Areva/EDF Pressurised Reactor with a generating capacity of 1,660 MW in a programme beginning with reactors at Flamanville in France, Olkiluoto in Finland (both under construction but experiencing long delays and cost increases), two under construction in China also delayed, and Hinkley Point in the UK, expected to be the first, though much delayed, plant in the UK's new nuclear programme (Environment Agency, 2010; HSE, 2009). A similar approach to standardisation has been applied to the design of reprocessing facilities at La Hague and in the surface disposal concept for the management of low level radioactive wastes implemented at both Centre de la Manche and Centre de l'Aube.

Hitherto, the French nuclear industry has been able to rely on the automatic and largely unquestioning support of the French state. The characteristics of a highly centralised and elitist organisational structure, commanding a ruthlessly efficient bureaucratic corps, provide firm political support for a nuclear industry which, in its turn, exemplifies French pride in technological advance. The industry is effectively a powerful arm of the state. By contrast, the institutions of local government and some parts of civil society which, in a democracy, might encourage pluralism and foster opposition have quite often proved quite biddable in France. This leaves space for the anti-nuclear movement in France to adopt a confrontational approach. Both in La Hague and in the proposed repository sites, local governments have proved, on the whole, receptive to the industry's economic inducements.

The French nuclear industry is conventionally portrayed as the archetype of technocracy. It represents the melding of centres of excellence, training scientists, engineers and civil servants (many trained at the grandes écoles) with the development of high technology and its application through an elitist governmental and administrative structure. French technocracy has both practical application and symbolic purpose (Ridley, 2006). It represents the aspiration of the French state to achieve technical and scientific excellence and achievement that promotes the image of France and defends its integrity. Gabrielle Hecht uses the idea of 'radiance' as a metaphor for the impression of the nuclear industry on French consciousness. It represents modernity, the idea of technology as saviour, redeemer and liberator (Hecht, 1998). In its early years the nuclear industry performed a spectacular demonstration of a country overcoming the defeats of war and austerity and moving forward as a leader in technological innovation and progress. 'Technological development was thus a tremendous spectacle, a drama propelled by scientists and engineers, and a display of national radiance' (ibid., p.209). In a similar vein, Sovacool and Valentine (2012, p.83) identify a kind of 'nucleocracy' or pantheon consisting of,

> the government which assumes the mantle of the gods with almost complete power, large industries and utilities closely aligned with the gods that serve as titans, and the general public who are the mere mortals that have little or no control over the country's nuclear future.

Thus France successfully promoted the image of a progressive and developing nuclear industry with clear goals and a firmness of purpose. This patriotic self-confidence and pride in its nuclear mission was a matter of envy to foreign observers who routinely encountered technical problems and opposition to their nuclear ambitions. For example, reporting on a visit to France in 1993 the UK government's Radioactive Waste Management Advisory Committee of which I was a member, reported with almost fulsome enthusiasm,

> The clarity of purpose that flows from this national approach creates an air of efficiency. The sense of common, shared purpose, makes it almost impossible to expose differences between Government and the various components of the industry. The result is that managers' competence and confidence is not undermined and there are public relations benefits. Overall it has to be said that the benefits stem from a Government prepared to identify and tackle strategic issues and to take difficult decisions.
>
> *(RWMAC, 1994, pp.3–4)*

A perceived necessity

This favourable profile of the industry has been maintained until recently by an assiduous attention to public relations. It is the case that the nuclear industry in France appears to have enjoyed widespread, if not enthusiastic, support among the population at large, though opinion polls provide a rather equivocal picture. For example, a poll taken at the end of 1997 revealed that over two-thirds of the sampled population felt it was not necessary to expand the industry but to maintain existing capacity. Of the remainder, 11% favoured further construction while 15% favoured immediate shut-down (Sortir du Nucléaire, 1998). This suggests nearly two decades ago an attitude towards the industry of 'perceived necessity' rooted in the dominance of nuclear energy in electricity supply. In the following decade a Eurobarometer poll indicated that over half (52%) of the French population were in favour of energy produced by nuclear power stations, with 41% against, and the remaining 7% undecided. While this favourable view was well above the average of 37%, France ranked only seventh of the 25 European countries surveyed (Eurobarometer, 2005). The polling evidence suggests a gradual improvement in nuclear's showing during the first decade of the century and a Eurobarometer poll published in 2010 indicated 45% in France favouring maintaining the proportion of nuclear energy which, added to the 12% favouring an increase in nuclear's role, outweighed those (37%) favouring a reduction (Eurobarometer, 2010). The Fukushima-Daichi nuclear disaster in Japan in March 2011 caused a sudden and significant downturn in popular support for nuclear energy in France as elsewhere. Indeed, a BBC poll taken in eight countries in the aftermath of the disaster found that 83% in France were against nuclear new build, behind Germany (90% against) but level with Russia and Japan. The curious exception was the UK where support for new nuclear had continued to rise from 33% in 2005 to 37% in 2011 (BBC, 2011).

From what can be gleaned from opinion poll evidence in the aftermath of Fukushima, the immediate downturn in support for nuclear energy in France has been followed by a slight recovery. But, the polling evidence is ephemeral, difficult to interpret and open to tendentious interpretation. For example, a Foratom report claimed that though 55% of the population in 2012 thought the risk of severe nuclear accidents was high, 'they do trust national authorities with controlling and ensuring the safe operation of nuclear reactors' (Foratom, 2014, p.7). Similarly, another survey finding 40% in favour of maintaining nuclear power claimed that, after some hesitant and inconsequential debate, 'the French seem to be back on the nuclear bandwagon. They are not ready to bury their pride, well aware that life would be more difficult without nuclear energy; their needs thus outweighing their fears' (The Typewriter, 2013). So, the ambivalence persists with, according to one analysis, a substantial proportion of the population (certainly over half) either undecided or against the use of nuclear energy in France although, perplexingly, a majority would maintain nuclear's share of the energy mix (WNN, 2013). This is contradicted by other polls which suggest a majority support a phase-out of nuclear energy in favour of an energy transition towards renewables (Schneider, 2013; IFOP, 2011).

The future of French nuclear energy has become a lot more uncertain both as a result of Fukushima and the election of President Hollande in 2012 on a platform which included the gradual reduction of nuclear's proportion of electricity supply from three-quarters to a half of the total by 2025. This led to an eight month national debate dealing with all aspects of energy policy as a prelude to legislation enshrining the energy transition in 2014. Meanwhile the development of a third generation had run into trouble with the first reactor at Flamanville costing twice as much and with its completion expected to be at least four years overdue. The industry had also experienced big cost increases and technical problems and disputes with the regulators with its development of new reactors at Okiluoto in Finland (reduced to one and estimated completion date 2018). Together with the failure to secure export markets and the long delays and financial problems encountered with its projected reactor at Hinkley Point in the UK the French nuclear industry had lost much of its former radiance.

The general public and political support enjoyed by the nuclear industry over many years has been reflected in those communities where nuclear power stations and other facilities have been developed. At the political level, nuclear waste policy is transmitted in relatively straightforward ways from central government through the representative institutions at regional, departmental and communal level. At the local level, support has been actively encouraged (opponents would say bought)

FIGURE 4.6 Nuclear reactors at Flamanville

Source: Électricité de France

by a range of financial incentives. These were offered by EDF during the period of power plant construction in the 1970s and 1980s. Quite aside from the investment and jobs a nuclear plant brings to the local community, local construction workers are favoured, local business is stimulated, electricity tariffs have been reduced and there are direct tax and rating levies which accrue to the local area and the department. Through skilful public relations the community benefits, in terms of facilities and infrastructure, are stressed; care is taken to minimise the physical impact of development (for example, by reducing the height of cooling towers or by concealing power stations in the cliffs as at Paluel and Flamanville). This policy of economic and community development has also been applied in those communities hosting radioactive waste facilities. It was the key to the successful establishment of the low level waste repository at Centre de l'Aube. I visited the site before it opened in 1994 with fellow members of the RWMAC and we reported favourably on the approach. 'The L'Aube facility today appears to enjoy local support on account of the employment it brings and as a result of the conscious effort to secure good community relations through liaison, economic "spin-offs", and social and environmental benefits' (RWMAC, 1994, p.5). A similar approach has reaped rewards both for the industry and the local community at La Hague and has been employed at Bure as part of the effort to secure a footing in the local community. While a rhetoric of public responsibility suggests a spirit of altruism and self-sacrifice, the reality is that communities expect, and receive, substantial inducements as part of the deal. The need for financial incentives has been built into the legislation indicating a commitment to compensation for imposing a potential risk. 'A less charitable view, of course, might see in such a policy little more than the crude buying-off of long-term uncertainties with a seductive coinage of more immediate but more superficial value than a nuclear-free countryside' (Boyle and Robinson, 1987, p.78).

Co-option or confrontation – nuclear opposition in France

The ascendancy of the nuclear industry in France and the public support it has secured has been a matter for admiration by foreign observers whatever their position on the nuclear issue. Conversely, the trenchant, determined and persistent opposition to nuclear policies and projects within France has often been disregarded and discounted for its failure to make any significant impression on the deployment of nuclear facilities. In the struggle between the industry and its opponents, 'the confrontations meeting the treatment of every event and project do not appear to be causing any major swaying in French nuclear policy, rather described as being extremely firm, especially in the foreign media' (Cézanne-Bert and Chateauraynaud, 2010, p.71). It is not quite so simple since anti-nuclear protests have scored some apparent successes which have influenced the course of radioactive waste management policy in France. Nonetheless, put simply, in the French context of power relations the discourse of Danger and Distrust which is strongly promoted through the anti-nuclear movement has encountered a formidable adversary in the dominant discourse of Trust in Technology.

Part of the explanation for the apparently unsuccessful anti-nuclear opposition lies in the way conflict is politically managed. Local participation in policy making has been institutionalised through local committees and confined to elected representatives and stakeholders rather than the wider public. In this way some of the potential opposition has been co-opted into the decision making process, giving it some influence but little power. Consequently, the more radical anti-nuclear opposition, excluded from participation, tends to be confrontational (occasionally violent) typically manifesting itself in large demonstrations and one-off events coordinated through national groups or networks. This reflects both the tendency for a central and national focus in nuclear politics and the fact that, locally, both people and action are thin on the ground. Even so, strong, and sometimes successful, local protests have been mobilised, notably against radioactive waste proposals, as we shall see.

National environmental groups concerned with the nuclear industry tend to have declined in importance in common with the experience in several other countries. Greenpeace France is small by comparison with its national equivalent in Germany. Greenpeace International, acting quite independently to prevent trade in nuclear materials, has in the past come into conflict with national groups concerned to ensure the repatriation of foreign wastes. Other national groups provide information for consciousness raising, publish newsletters, alert the media and carry out occasional actions but, overall, their resources are limited and overstretched and devoted to a range of issues. An example is Robin des Bois, whom I interviewed in 1996, which was active in lobbying on nuclear issues such as reprocessing, waste transfers and new nuclear power in France but later devoted more attention to international environmental issues such as the ivory trade. The World Information Service on Energy (WISE), based in Paris, has sustained its presence on the nuclear front producing detailed critiques of the nuclear industry and thereby providing some significant co-ordination of an anti-nuclear analysis but it is not engaged in the organisation of action at grass roots level. This tends to be provided by local action groups which develop when a specific threat is presented but fade away when the threat is removed or the development becomes irresistible. The most effective organisation, the network *Sortir du Nucléaire,* has a membership of over 900 organisations and almost 60,000 subscribers (Sortir du Nucléaire website, www.sortirdunucleaire.org). As its name implies its policy is to achieve the phasing out of nuclear energy and it devotes its resources to the multifarious tasks of an active NGO – supporting actions, gathering petitions, providing information, raising consciousness, lobbying decision makers and generating publicity. Overall, the French anti-nuclear movement appears characterised by diversity, fragmentation and intermittency that are a product of its history. On the other hand, especially when modernisation threatens cherished traditional communities, it is also capable of mass demonstrations incorporating not merely anti-nuclear activists but also farmers, citizens and other groups in a cross-cutting combination to defend their territory.

FIGURE 4.7 Clashes between protesters and police at the Superphénix fast breeder reactor at Creys-Malville in July 1977

Source: hinifoto.de/akw/malville.jpg

As the nuclear industry was rapidly expanding during the 1970s and beyond, a discourse of Danger and Distrust was evident in relations between the industry and its opponents. In these early years the French anti-nuclear movement was powerful and effective, able to mobilise mass demonstrations. It is sometimes forgotten that France was the scene of some of the biggest and most violent protests culminating at Creys-Malville, the site of the Superphénix project in 1977 where 80,000 gathered and one protester died with others badly hurt. In Brittany, opponents secured the support of François Mitterand who, campaigning for the presidency, cancelled the proposed power station at Plogoff in 1981. Once the industry had become firmly established there were fewer targets on which to focus an attack. In any case the violence experienced at Creys-Malville and other sites where resistance had 'been bludgeoned into disarray by the crude exercise of State authority' (Boyle and Robinson, 1987, p.76) had demoralised many activists. Nonetheless, it was possible to claim some signal triumphs such as the rejection of Plogoff and the abandonment of the Superphénix though these might have occurred anyway. La Hague, too, became enveloped in this early phase of conflict as reprocessing developed and before the plants became an established part of the community.

By the 1990s, as the nuclear power programme was being fulfilled, attention turned more towards the rear end of the nuclear cycle and especially to the problem of finding a site for a deep geological repository. As attention switched to managing the nuclear legacy so a discourse of Cooperation and Consensus began to develop. While the French nuclear industry at the end of the twentieth century still possessed enormous resources of power, it was also showing some signs of vulnerability as the issues of reprocessing and waste management occupied greater attention. This shift in both focus and discourse can be traced and understood in the local context of the two communities that are the subject of the rest of this chapter.

La Hague – nuclear's past and future

The heart of the nuclear enterprise

Over the years since it was first begun in the 1960s, La Hague has represented the French determination to develop the full panoply of nuclear processes. Unlike Sellafield where, as we saw in the previous chapter, reprocessing is being phased out, La Hague is very much in business as the reprocessing centre around which the French nuclear industry revolves. It is worth, therefore, considering why it has managed to achieve and sustain such a dominating role.

Altogether the La Hague nuclear complex covers around 300 hectares, an area 3km by 1km. Its impact on the local area, physically, economically and socially, is profound and, both in national and, to some extent, international terms, La Hague is the political focus of the French nuclear industrial enterprise. Initially, it had a military function as a backup should there be a problem with the first reprocessing plant (UP1) at Marcoule in the south in the Rhône Valley. But the primary function of the UP2 plant at La Hague was for reprocessing spent fuel, initially foreign, and then French, and in 1994 it was expanded to a capacity of 800 tonnes per year. A second plant (UP3) of similar capacity was commissioned in 1990 and dedicated to reprocessing foreign fuels. By 2013 with foreign fuel reprocessing virtually over, the two plants had excess capacity and their combined strategic capacity had been reduced to 1,000 tonnes. Apart from the two reprocessing works, the La Hague complex also includes waste conditioning and storage facilities. Nearby is the low level waste surface disposal facility, Centre de la Manche, which was decommissioned in 1994 when the new facility, Centre de l'Aube, opened up in eastern France.

While reprocessing has been prevented, abandoned or challenged in many other countries, France, and to a lesser extent the UK, has promoted it as the solution to the environmental, safety, security and waste management problems of the nuclear industry. Reprocessing has military origins in the production of plutonium for weapons at Marcoule. Later, it was assumed that fast breeder technology would usher in a new nuclear era dependent on plutonium. With the ending of the Cold War and the shut-down of fast breeders, including Superphénix, the case for reprocessing appeared fatally weakened. Plutonium, the key product from reprocessing, had become a dangerous material without an apparent use. The French response was to find a use in Mixed Oxide (MOX) fuel fabricated in the Melox plant at Marcoule. By recycling plutonium, its advocates claimed, not only would it be a resource, it would also be easier to manage by minimising waste production and reducing the danger of proliferation. The advocates of 'moxification' pronounced this as a challenge and an opportunity.

A challenge because poor management of this material would maintain a problematic situation in terms of proliferation; an opportunity because such plutonium represents a high value energy source that the civilian industry is capable of using efficiently, actually turning it 'from swords to ploughshares'.

(Bastard and McMurphy, 1995)

Above all, MOX was capable of 'providing an answer to the problem of surplus stocks of strategic military-grade plutonium' (COGEMA, 1997, p.25).

Reprocessing has led a seemingly charmed life for its advocates by continually reinventing its purpose. EDF (Electricité de France), the monopoly electricity supplier, provides the demand for plutonium recycled into MOX fuel to power 24 of its 58 reactors. Having signed the contracts it is not surprising the company justified its choice of reprocessing. 'It is an economic choice – a choice for the future' (interview with Jean-Pierre Chaussade, EDF, 2004). But, EDF's demands will not absorb all the plutonium produced and foreign contracts for MOX are unlikely to fill the gap so long as uranium fuel is in abundant supply, leaving the industry, chameleon-like, to find yet a further purpose for its continuing existence in the promise of fuelling the fourth generation of nuclear reactors which, if they appear at all, will not be deployed anywhere any time soon. Although, in theory, promising to reduce or eliminate many of the disadvantages of current nuclear technologies, notably waste volumes and radiotoxicity, in practice, much research and experimentation is necessary before commercial development of new reactors can begin at around 2040 at the earliest (Sovacool and Valentine, 2012). Nevertheless, optimistic scenarios are presented for 2040 indicating further PWR reactors to recycle about 15% of the plutonium combined with a series of Fast Neutron (breeder) Reactors (FNR) to ensure multi-recycling of about 172 tonnes of plutonium derived from MOX by that date, expanding as more plutonium is produced (CNE, 2012). At a time when France is having problems with its existing nuclear programme at Flamanville, when much research remains to be done to demonstrate feasibility of projects and when the country is reducing its nuclear programme, the prospects for such developments may appear in the realm of fantasy.

In the meantime, spent fuel is accumulating both at reactor sites and at La Hague where storage capacity is being increased to accommodate 17,600 tons (IPFM, 2011). The future of this fuel is unclear since it may not all be reprocessed but there are no plans for its direct disposal. Similarly, MOX fuel, once used, is not for the present being reprocessed and its status is ambiguous since, until its future is clarified, it is neither a resource nor a waste. There are no plans for disposal of MOX which generates much more heat than low-enriched uranium fuel and so requires longer for cooling and more space in a repository. It all depends on future nuclear strategy, whether France embarks on a new generation of nuclear reactors fuelled by plutonium as currently assumed or significantly reduces its commitment to nuclear energy which is the preferred strategy in the policy shift towards renewables. In any event, and especially if nuclear is in decline, the accumulation of spent fuel and the declining demand for MOX will intensify the problem of managing wastes for which there is no long-term solution. The geological repository proposed for Bure is designed for vitrified high level wastes and conditioned intermediate level wastes. At present there is no designated long-term management route for MOX, unreprocessed spent fuel, surplus plutonium and some other wastes, beyond continued storage which makes La Hague, *de facto,* the primary location for the long-term management of both high level wastes pending disposal and those for which no disposal route is available as well as a large volume of sludges, and other wastes from earlier operations awaiting conditioning. Meanwhile, it has become increasingly evident that commercial reprocessing 'although originally introduced to obtain plutonium fuel for starting up fast-neutron reactors, is now clearly established as the national policy for spent-fuel management' (IPFM, 2011, p.33).

For the present, reprocessing has survived at the heart of the French nuclear cycle partly through a remarkable capacity for adaptation, and partly through policy inertia. According to Philippe Pradel of COGEMA, plutonium, a valuable and reusable resource, must be conserved (interview, 1999). But recycling ultimately produces wastes that have to be managed. This waste is vitrified which, according to the industry, presents the safest method of storage and disposal. Indeed, with recourse to hyperbole, La Hague was described by its former management as the 'safest plant in the world, the most controlled plant in the world' (interview, Patrick Fauchon, COGEMA, 1996), a claim that has not gone unchallenged by nuclear opponents (see especially Guillemette and Zerbib, 2012; Guillemette, 2012). In its latest manifestation, then, as the method for managing wastes, it is claimed that reprocessing saves natural uranium, reduces radiotoxicity in waste streams and minimises waste volumes thereby reducing costs (Giraud and de Lepine, 1995). While these assumptions have been strenuously challenged on all counts, the point is that the French industry exhibits a mixture of confidence and vulnerability in its strident assertion of the merits of reprocessing. This is no less than a defensive strategy designed to protect its power and its markets.

The survival of reprocessing in defiance of technical, commercial and social realities continues to baffle its critics. Ben Cramer, a writer and journalist, commented during an interview in 2004; 'La Hague is a crazy phenomenon. It is magic. We thought reprocessing was finished because of shipments and proliferation but it has survived. We said MOX had no future but now 20 reactors have it.' This idea that the fixation with reprocessing can only be explained by appeal to the supernatural is echoed by André Guillemette, a former worker at the Arsenal who has subsequently devoted his life to revealing the problem of emissions and discharge from the La Hague plant.

> Reprocessing is their religion. For me, reprocessing is nonsense – nobody knows what it will cost, the plutonium is unused. Although reprocessing is justified by MOX, in truth La Hague is really a storage site. It continues for a new generation of reactors.
>
> *(Interview, 2013)*

Yves Marignac of WISE, Paris, a long-time critic of the policy, recognised the contradictions years ago. 'Nothing much changes. But it's like opening Pandora's box – the whole logical construction falls apart. The more the reality becomes different to what you want to believe, the more difficult it is to recognise it' (interview, 2004). This apparent blindness persists nearly a decade later. 'It is a feeling that pervades not only the government and managers of the nuclear industry but affects the workforce and local community, too' (interview, 2013). Frédéric Merillier of Greenpeace noted the sense of denial when I interviewed him in 2004. 'There is no debate at local or national level about La Hague – they don't want to ask the question because it looks as if it would stop. They don't want to look at the future'. The problem with policy drift was spelled out by WISE: 'If the fundamental principles that have precipitated the French nuclear industry into its current crisis are not acknowledged, much effort and money will be invested in an approach which is doomed to fail' (WISE, 2015, p.5). But, it is the combination of denial, inertia and adaptation that has kept reprocessing at La Hague in business and which promises to sustain its existence for many years to come.

Conflict and coexistence

So, La Hague has continued more or less the same by adaptation to changing circumstances. As Yves Marignac puts it: 'Although less and less rational, La Hague is still the core of French nuclear strategy' (interview, 2013). It is a classic case of the persistence of nuclear activities in a peripheral community. Like Hanford (Chapter 2) and Sellafield (Chapter 3) La Hague is at the core of the plutonium economy, no longer engaged in the military sphere but maintaining its foothold simply because it is there, with a continuing legacy from the twentieth century that must still be managed in the twenty-first. And, like Sellafield and Hanford, and many other nuclear communities, La Hague owes its existence to a combination of geographical, political and social circumstances responsible for its identification, ideology and identity.

La Hague was originally selected for a combination of technical and political reasons. The strong winds and powerful offshore current around Cap de la Hague ensured dispersal and dilution of gaseous and liquid effluents from the plant. It was in an economically backward area where regeneration would be welcomed. According to Patrick Fauchon of COGEMA, it was 'out in the wilds', a place open to incoming development. Moreover, it was considered sufficiently remote for the secretive nuclear industry to insinuate itself with little resistance. As Didier Anger, a long-term opponent of nuclear activity in the region and the leader and inspiration behind the protest group CRILAN,[4] expressed it to me; the site was 'au bout du monde' (interview, 2005). In short, it fulfilled the basic criteria associated with 'peripheral' locations which I set out in chapter 1 (Blowers and Leroy, 1994). La Hague was an intrusion of the modern in a traditional society, an exemplar of the juxtaposition of opposing elements that create tensions predominantly latent but occasionally giving rise to more overt conflicts. La Hague evokes the uneasy settlement that was initially secured by the nuclear industry in this remote corner of France leading on over time to a more rooted existence as an integral and inescapable element in the make up of the community. Its history and geography offer an explanation for the development of an industry that is as entrenched as anywhere in the Western world.

When the industry arrived, the Cotentin Peninsula was a backward region. It is a land of poor soil, gorse covered moor, yellow and green with the sea pounding the steep cliffs and wide sandy bays. Françoise Zonabend's description at the head of this chapter is evocative of its landscape (1993, p.13). Its essence is the moor where 'gorse has slowly, relentlessly, conquered the leafless underbrush, swallowing, exhausting the soil with lethal gluttony' (Hochman and Decoin, 1997, p.12). It has an austere beauty with the ever-changing weather shifting between grey skies and persistent drizzle, to fresh winds, white clouds and the sun's rays dazzling the big seas beyond. It is a harsh, unyielding land mainly given over to grazing and supporting small, grey villages on the uplands and around the bays along the coast. Farming and fishing are the traditional occupations. Cherbourg, the largest town, has a massive harbour too big for the ferry trade it now supports. The nuclear industry is a prominent presence, from the large hanger of the Arsenal in Cherbourg, to the reprocessing works perched prominently on the top of the moor, to the Flamanville power stations blasted into the cliffs on the western shore.

FIGURE 4.8 Coastal and rural landscape around La Hague, in the Cotentin Peninsula

Source: Author

La Hague's arrival in this furthest shore of France did not go altogether unchallenged but, over time, the plant has become integrated into both landscape and community. The works at La Hague emerged into an area with very little social mobilisation, a region where peasants were not very well informed and generally accepting, willing to be paid for land that was worth very little (interview with Didier Anger, 2013). In the early years it was the nuclear power plant at Flamanville that attracted most protest at a period when opposition to the nuclear industry in France was more vigorous than it has ever been since. The course of this conflict was described to me by Didier Anger in 1996 during the first of three conversations I have had with him over the past two decades. At Flamanville, protests began in 1975 and were on a larger scale than at Plogoff in Brittany. During 1977 around 8,000 people gathered and the site was occupied for a month followed by a legal action which, though it denied a permit, did not succeed in preventing construction. The Plogoff decision redeemed an election promise though in the comparable case of Flamanville, according to Didier Anger, the local community felt betrayed by the politicians (Anger, 1987).

At La Hague, too, there were demonstrations in 1972, and in 1976 the trade unions organised a strike against the privatisation of the plant. They were supported by a demonstration of 10,000 organised by CRILAN thus affirming a linkage between unions and environmentalists in concerns about the environment, safety and the expansion of the plant. In 1979, CRILAN mobilised around 7,000 demonstrators against the first arrival of spent fuel from Japan. But, like other protests, it was suppressed by strong police tactics. Another, larger protest of 25,000 occurred in 1980 against foreign reprocessing. Mitterand's election in 1981 heralded a less enthusiastic nuclear policy. The anti-nuclear movement anchored in the political left found its outright opposition to nuclear power compromised with the accession of a government rather more sympathetic to its aims. The government, too, found its ambitions on the nuclear front compromised. According to Didier Anger, while Mitterand cancelled Plogoff and blocked foreign contracts, the collapse of the franc made the abandonment of reprocessing politically impossible. In return for its support for the French currency, Germany, with little capacity to store its waste and prevented by opposition from developing its own reprocessing industry, continued to export its spent fuel for reprocessing to France. The agreement for the return of wastes from reprocessing to Germany became a source of protest in both countries.

The 1980s were a relatively quiescent period in terms of anti-nuclear protests in France. In part this reflected a decline in protests in France as a whole especially during the period of socialist government. Once the nuclear industry had become firmly established with its power stations all on line, there were few targets on which to focus an attack. In any case, the violence experienced at Creys-Malville and other sites had demoralised many activists. Although from time to time there have been further protests and mass demonstrations, protest has become more sporadic and focused on wider issues. By the 1990s, La Hague had become the focus of international attention with the return of plutonium and wastes to Japan and, later, Germany. The shipment of 250kg of plutonium from Cherbourg to Japan in 1992/3 aroused widespread international protest with the refusal by some countries to allow use of their sea lanes and a demonstration at the receiving port of Tokai. Two years later, a similar protest orchestrated by Greenpeace International and including a network of NGOs from many countries, dogged the voyage of the *Pacific Pintail* with its cargo of 14 tonnes of HLW bound for Rakkasho-mura in Japan (Blowers and Lowry, 1996). Cherbourg was a major focus with the attempt to prevent the sailing and, while at sea, the ship ran the gauntlet of protest including the Chilean navy which prevented it entering Chilean waters. The protest was against the carrying of dangerous cargoes across international waters but it also raised issues of nuclear policy bound up with reprocessing and the management of nuclear waste. The Japanese shipments created tensions within the environmental movement. Whereas Greenpeace International wished to prevent the shipments, French campaigners, notably local ecological groups, CRILAN and Greenpeace France, asserted the wastes must be returned in line with agreements and policy.

As was the case at Sellafield (Chapter 3), so at La Hague, the return of plutonium and wastes to Germany was controversial. On the one hand, it was generally agreed

FIGURE 4.9 Nuclear cargo ship *Oceanic Pintail,* renamed from *Pacific Pintail* in 2012

Source: International Nuclear Services

that these materials must be sent back. This was a provision in the contracts, a requirement under the French nuclear law of 1991 and was regarded on all sides as a moral and ethical necessity. As a trade unionist working at La Hague put it to me, 'Sending it back is a moral and ethical question that each one who profits must accept both the good and the bad' (interview, 1996). On the other hand, the return of German wastes encountered concerns about safe transportation. The discovery of a contaminated flask during transit in 1998 brought a temporary halt to the shipments. Meanwhile, in Germany, mass protests against shipments into the Gorleben store had delayed movements of waste and the election of a SPD/Green coalition government had opened up the whole question of reprocessing. From a French perspective, events in Germany appeared a temporary hiatus in the movement of nuclear materials but, at the same time, contributed to the aims of environmentalists to strike at reprocessing. In any event, the conflicts over the return of wastes indicated that La Hague, on the geographical periphery of France, remained at the political centre of the debate over the future of the nuclear industry. In particular, the problem of managing the wastes generated and stored at the plant had become a major political issue for the industry. Not only was there the problem encountered with returning foreign wastes, France increasingly had to turn its attention to what to do with the wastes generated from its own nuclear industry, much of which was accumulating at La Hague. Spent fuel from reactors around the country was transported to La Hague for reprocessing and plutonium was shipped from there to Marcoule to be made into MOX while the high level wastes remained in store awaiting possible transfer to a geological repository sometime within the next hundred years. Thus, much of France's nuclear cycle passes through La Hague at some point. La Hague, though peripheral in its geographical location, has become the core of the country's nuclear complex.

Once La Hague's pivotal role had become established, paradoxically it ceased to be the focus of protest. The arguments against reprocessing – the lack of markets, the surplus of plutonium, the build-up of wastes – had become familiar, if disregarded. Above all, there was the argument that France, far from benefiting from reprocessing, was bearing a disproportionate cost. WISE-Paris made the point back in 1994,

> The situation of the wastes from spent fuel reprocessing demonstrates a problem which until the present day has attracted little attention in France: the nation is *exporting* more and more nuclear services but is *importing* inherent risks of many types and in particular the problem of waste management.
>
> *(Homberg et al., 1994, p.23)*

In fact it is La Hague that is importing the risks from other parts of France as well as overseas.

Nearly 20 years on, reprocessing continues but in conditions of increasing costs and investment requirements. With the loss of practically all its overseas business it has become precariously dependent on EDF as its sole market for MOX. More and more La Hague is becoming focused on its function on reprocessing as a means of waste management. As the French nuclear industry shrinks so La Hague, in common with other parts of the nuclear sector, 'must urgently shift its focus to the maintenance of current reactors and decommissioning and nuclear waste management services' (WISE, 2015, p.5). In its diminished role, nonetheless La Hague continues, persistently and tenaciously, as the core on the periphery.

A precarious existence

By the turn of the century, then, La Hague had ceased to be a focus of mass protest as the reprocessing plant had become embedded in the local community. Attention in the Cotentin was, once again, drawn to the production of nuclear energy as it had been during the 1970s and, specifically, to the development of a new power plant at Flamanville. This project drew a mass protest at Cherbourg in 2006 and a large 'manifestation' against the EPR reactor more generally gathered at Rennes in Brittany in 2011. Opposition to La Hague had become less visible though local groups continued to try to expose the risks and dangers emanating from the complex. The number of protesters is today far smaller than the masses mobilised in the 1970s and they are linked to broader national and international movements, symptomatic of the internationalisation of nuclear conflict. Among the more local groups is ACRO,[5] an independent NGO based in Caen devoted to measuring, monitoring and publishing the levels of emissions from nuclear plants. But, with the policy on reprocessing settled, at least for the time being, La Hague was not politically controversial but had become the economic fulcrum for an entire sub-region. The industry could rely on its economic presence and power in an area with few other activities. Moreover, it was a declining area losing its manufacturing base including oil drilling platforms, textiles, electronics, naval construction, car distribution depot and port activities. By contrast, the nuclear industry had attracted an adventitious population earning more than those in traditional occupations, dependent on the industry and, in consequence, providing it with an increasingly strong basis of local support. The area attracts few other enterprises and the nuclear industry builds up the local tax base and provides a range of social facilities. It had encouraged the modernisation of the infrastructure with a four-lane highway right up to the plant and an electrified railway to Paris. With its domination of the economic landscape there is no 'countervailing power to that automatically and somewhat mysteriously exerted by COGEMA purely by virtue of its dominance over the regional economy' (Zonabend, 1993, p.60). Though peripheral it is relatively prosperous.

Over time, relationships have matured to the point where nuclear is an integral, it might almost be said, traditional part of the economic landscape in the northern Cotentin. Even 20 years ago, one local manager, Philippe Fournier, was able to claim, 'The general relationship is getting better and better but slowly' (interview, 1996). La Hague, like other nuclear plants in the Western world, has responded to the requirements of the 'deliberative turn'. A policy of openness, information

FIGURE 4.10 Anti-nuclear protest posters

provision and participation is *de rigeur*, an obligatory element of the industry's integration and engagement in its local region.

As in every French nuclear installation there is a local information commission (in this case the Commission Locale d'Information or CLI) which mediates with the public providing information including environmental monitoring. A newsletter is distributed to all homes in the department three times a year and Areva takes a proactive approach to public relations and communication. The company has considerable economic impact as the largest employer in the north Cotentin with around 5,000 employees, with over €400 million worth of purchasing from suppliers in the region, paying €80 million in taxes and investing over €100 million a year. Through the organisation of 'Nucleopolis', Areva supports investment, research and training to contribute to diversification and development in the region. The company exudes a sense of pride, well-being and commitment to solidify its presence. 'We are here and we want to live here. We have to be proud of what we're doing. It's the basis of respect' (interview with Katherine Argant, Areva, 2013).

Yet there is a detectable ambiguity in the relations between the industry, its workforce and the local community that has persisted over the years. On the face of it, the jobs, wealth, taxes, social facilities and economic prosperity the industry brings to an underdeveloped part of France is the important thing. The point was echoed by one trade unionist I interviewed: 'The industry is not necessarily popular, but it is necessary... it would be a catastrophe if it closes. It is a bad necessity' (interview, 1996). This same sense of anxiety and inevitability is evident in Didier Anger's comment nearly two decades later.

> There are no complaints about the industry because the area needs it. The soup is good and we want more. Yet everyone is fearful of nuclear at the same time. They are stuck between fear of nuclear and fear of the economy. We are all *immediatistes*.
>
> *(Interview, 2013)*

The traditional and the modern

For the first two decades or so of its existence, La Hague was accommodated, if not wholly accepted, into this traditional community. Françoise Zonabend, researching during the 1980s in what she called 'this spot at the back of beyond' (Zonabend, 1993, p.ix) described its villages 'inhabited by new, heterogeneous, mobile populations with whom the old residents attempt to coexist or whom they pretend to ignore' (p.38). A worker in the industry for 22 years commented, 'I'm still a stranger here' (interview with trade unions, 1996). Yet, as Zonabend observed, the traditional was being gradually eroded in face of the modern. 'In this context of modernity, this high-tech atmosphere ushered in by the plant, people's links with the environment are loosening' (1993, p.46). Of course, there are other processes at work in transforming the society but the presence of the nuclear industry has reduced the isolation, it has accelerated the change. The ambiguity about risk, the repressed anxiety and the fatalistic adaptation are symptomatic of a peripheral community where power has shifted and where cultural norms are changing.

This ambiguity is reflected in the attitudes of workers, the industry, community leaders and anti-nuclear groups. Trade unionists working at the plant have participated in past actions and strikes and some were sympathetic to ecological concerns. While the Communist CGT remained implacably pro-nuclear, the CFDT[6] was more equivocal and opposed the reprocessing of foreign spent fuel. In the past the unions played a significant role in drafting clause 3 of the 1991 French nuclear waste law which states: 'Radioactive wastes from foreign sources shall not be disposed of in France, even if it was reprocessed on French territory, nor shall it be stored beyond what is customary and normal after reprocessing'.[7] They continue to argue for diversification of the local economy but nevertheless recognise the significance of the industry to the local economy and society.

Similarly, the local political leadership has tended to be supportive of an industry in which so many work and which brings material benefit to the region. As Yves Bonnet (in 1996 Gaullist Deputy in the French Parliament and President of the local liaison committee), referring to the economic impact of the plant, commented to me: 'It's difficult to say what people feel – economy is the important thing. I don't say it's good; it is a fact'. Anti-nuclear groups, now thin on the ground, focus their attention on the risks rather than the existence of the nuclear industry in the Cotentin. The aim of groups such as ACRO is to arouse awareness, provide an independent check and disseminate information. Their ambivalent relationship to the workers at COGEMA was neatly summed up by one of ACRO's scientists, 'Deep inside they think what we're doing is important. But, they are afraid for their families and themselves' (David Boilley, interview, 1996).

There is, deep down, a recognition that, in Zonabend's words, 'with nuclear power, man, like the sorcerer's apprentice, has started a process that he cannot stop' (Zonabend, 1993, p.128). It induces passive acceptance of the threat, anaesthetises concern for the future and induces in the population a reluctance to confront the present realities. Once again, Didier Anger (interview, 2005) provides an apt description of this collective comatose condition. 'Le Cotentin ressemble à l'autruche: elle met la tête dans le sable, elle ne voit pas le chasseur, mais le chasseur lui tire dans les fesses avec son fusil.'[8] As La Hague has settled more comfortably into its surrounding region, so attention has switched to the wider, in space and time, problem of managing the nuclear wastes that are produced from reprocessing and which are accumulating at La Hague and elsewhere. In particular, attention has focused on an area on the other side of France and on a small village and its surrounding territory selected as the site for the deep geological disposal facility for the permanent emplacement of the country's highly active radioactive wastes.

Bure – deep disposal in la France profonde

In search of a site

Bure emerged as the preferred site for France's deep repository during the 1990s. For some time before, the search for a site had been beset with all the uncertainties, controversies and impediments experienced elsewhere. In its initial search for potential sites for a deep level repository for high level wastes, France, like Britain and the United States, identified technically suitable sites but found it had to retreat in the face of local opposition. During the 1980s the Castaing Commission proposed that two sites be chosen for investigation, with one to be selected for the deep repository. In the event, in 1987 four sites, each with different rock characteristics, were identified for exploration as potential repositories. According to Claire Mays, despite some initial soundings, the sites appeared to be announced without warning or negotiation, a classical case of Decide Announce Defend arousing opposition in each area (Mays, 2004). They were in the departments of Aisne (clay) in the northeast; Deux-Sèvres (granite) and Maine-et-Loire (schist) in western France; and Ain (a salt formation) in the east (Figure 4.10). Christian Bataille, the deputy from the department of Nord who became a key figure in the subsequent development of radioactive waste policy, described the reactions of the local populations in his report to the National Assembly. Once investigations had begun at the sites, 'firm and resolute opposition from a large proportion of the population concerned' was encountered (Bataille, 1990, p.11). Indeed, in two cases, 'this opposition was of such a scale that serious public disorder was only avoided by the firm implementation of legal measures' (p.11). The public reaction at the sites was, according to Bataille, 'evidence of a more general malaise which is gradually spreading amongst our compatriots who are confronted with a technology of which they do not really understand all the ins and outs' (p.13). As a result, the programme had to be abandoned in February 1990. It became clear that, if a policy were to succeed, it would have to address social and political concerns as well as technical criteria. Up to this point site selection had been conceived as mainly a technical problem, finding somewhere with suitable geology for the presumed solution of deep geological disposal. This 'technological problematisation' created, according to Barthe, a 'path dependency' or 'lock in' 'which finally appeared as a condition of the possibility of change characterising the way in which the issue was addressed' (Barthe, 2009, p.946). In other words, the technical solution now became a political problem.

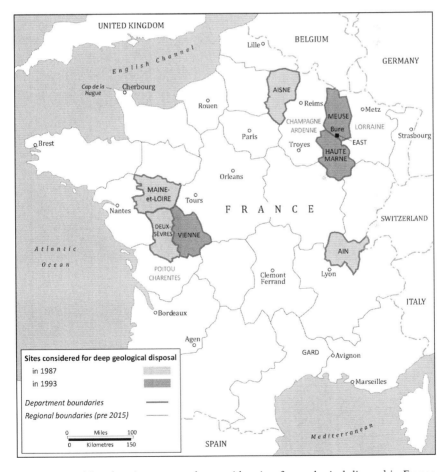

FIGURE 4.11   Map showing areas under consideration for geological disposal in France in 1987 and 1993

Map by: John Hunt

The government's response was to provide a combination of economic incentives, dialogue and negotiation enshrined in a legislative package. The Law of 30 December 1991 on Research in Radioactive Waste Management (French Republic, 1991) thereafter known as the 'Bataille Law' after its progenitor, marks the transition from the scientific and technical approach to a socially responsive decision model (Mays and Poumadère, 1996). It was the first piece of comprehensive national legislation for the nuclear industry in France. Its key principle, reflecting sustainable development, was that 'High-level, long-lived waste shall be managed in a manner that ensures the protection of nature, of the environment and of public health, and that respects the rights of posterity' (Article 1). Article 3 ruled out disposal of foreign wastes on French soil. The new policy was set out in Article 4 which identified three alternative areas of research or axes: one into partitioning and transmutation (P and T) of long-lived radionuclides; one into evaluating deep disposal options through creating underground laboratories; and a third into long-term surface storage techniques. The CEA was to be the lead organisation for developing the first and Andra for the other two axes. The progress of the research would be evaluated every year by the national assessment board, the CNE,[9] composed of scientific and engineering experts. These programmes were to be conducted in parallel and would be evaluated over a period of no more than 15 years resulting in a decision by Parliament in 2006 which would give the go-ahead for a deep disposal facility if it were the preferred option. Other provisions, relating specifically to the development of underground laboratories, emphasised the need for public involvement and safeguarding the public interest. Under Article 13, Andra was to assist in the development of the research programmes and to manage the long-term disposal facilities. A public interest group (Article 12) would 'undertake companion [i.e. economic support] activities and ... manage local resources to benefit and facilitate the construction and operation of each laboratory'. Finally, local information and oversight committees for each underground laboratory would be consulted on all matters affecting the local environment (Article 14).

Although the Act insisted on three axes of research, it was clear that finding sites for underground research laboratories, one of which might eventually become the location for a deep disposal repository, was likely to prove the most politically sensitive issue. Accordingly, Christian Bataille who had recommended the new approach, was appointed mediator, charged with 'leading public involvement prior to the selection of sites'. The process was to be conducted in accordance with principles of 'openness, concern for future generations, and the obligation for safety and environmental considerations'. Bataille recognised that, in order to overcome 'watchful suspiciousness' there was a need to shift from 'the cult of secrecy and self-satisfaction' to a process of 'political and social consensus' (Bataille, 1993, pp.3/4). His 'Mission' to find sites would reflect the ideals of openness and democracy, 'giving audience to the public's elected representatives and those of its different socio-economic groups'. The approach adopted was to call for expressions of interest, in other words for volunteers. But, in order to stimulate interest, an attractive package of incentives was also on offer. Apart from the substantial capital investment and employment created by the laboratory there would be FF60 million per year (equivalent to €9 million) to be spent on economic incentives for the local region. This was not to 'buy consciences' since, Bataille argued, 'Such criticism is obviously an insult to the sense of civic duty expressed in certain Departments' (1993, p.23). Indeed, the incentives were envisaged as a stimulus to economic development in the areas surrounding the sites.

In pursuing the need for a political and social consensus, the Mission organised public hearings in those departments that had expressed an interest. There was an emphasis on openness, involvement and the provision of detailed information. The process involved elected representatives who 'exhibited a noticeable spirit of civic responsibility' (ibid., p.16), chambers of commerce and regional associations (interested in the economic incentives), trade unions (on the whole unwilling to take a position) and environmental groups (whose views varied but were, on the whole, opposed to the project). The mixed reactions were predictable with some departments more politicised over the issue than others, giving rise to defensive reactions. In order to build the trust essential to success, the Mission emphasised the need for information and especially the provision of support for elected representatives who 'will have an essential role to play in the process leading to site selection' (p.21). It is clear that the approach relied heavily on achieving results through the political processes in which local politicians and especially mayors provide the support and departmental and regional councils the legitimation for decisions enacted by the state. While this is, formally speaking, a democratic process, it remains somewhat elitist and exclusive. Participation by the public, though encouraged under the Act, tends to be seen as a means of gauging reactions to proposals in the search for sufficient consensus. But, political support is not the same thing as consensus which is altogether more elusive. 'The consensus forms out of composite, sometimes unstable elements, which are alternately political, social, economic or cultural in nature' (p.26).

In the search for sites for laboratories it was clear that the four departments chosen during the 1980s must be ruled out since opposition there still persisted. The search for volunteers did, however, achieve results (CoRWM, 2008). Altogether 30 requests for information were received. Following a screening out of unsuitable geologies by the Bureau de Recherches Géologiques et Minières (BRGM) the search was narrowed down to eight departments as potentially suitable in geological terms. In the event the Mission rejected two when opposition developed and two where it was felt more time would be needed to ensure positive interest. This left four: Gard (in the south where Marcoule is located); Vienne (around Poitiers in west-central France); and Meuse and Haute-Marne which were joined into a single site, called the East site in 1995 (Figure 4.10). The population around these sites exhibited 'consensual elements' and their candidacy reflected a 'unique value of commitment and a uniting force' (Bataille, 1993, pp.30/31). The level of support was confirmed in votes by elected representatives in each of the three areas in 1997 (Boissac and Tamborini, 1999). At the most local level, the commune, votes were overwhelmingly in favour in all three areas. At the departmental level votes were again highly favourable. However, at the regional level, where the economic incentives would have less impact and where a wider constituency of protest might be expected, the results were less favourable. While a favourable majority was achieved in Poitou-Charente (within which Vienne is located), the East site was split between a majority in favour in Champagne-Ardennes (containing Haute-Marne) and a majority against in Lorraine (in which Meuse is situated). In the southern region around Gard the regional councils were overwhelmingly hostile. However, it is indicative of French attitudes towards the nuclear industry that, at the local level at least, democratic bodies are prepared to vote in effect in favour of the possibility of a deep geological repository.

Having passed an important political test the sites had also to be confirmed on technical grounds. Investigation of the sites by Andra was undertaken between 1994 and 1996. The eastern site, located at Bure in Meuse near the border with Haute-Marne, an extensive clay formation with promising hydrogeology and stable geology, was found to display 'no prohibitive feature in terms of siting criteria' (Andra, 1997a, p.20). The granite site at la Chapelle-Bâton in Vienne revealed potential problems of fracturing in the rock and seismic activity. This required analysis of long-term geological and climate changes relative to the integrity of the waste containment suggesting that the results were insufficient to make a safety case for a repository without further investigation (Andra, 1997b). In the case of the silt site at Marcoule in Gard, faulting, permeability in surrounding structures and the possibility of a major incursion through land and sea changes led to a tentative assessment (Andra, 1997c). In any case, local opposition, especially from the local wine industry, compromised the favourable reaction from a nuclear oasis dominated by the Marcoule reprocessing and fuel fabrication complex. The analysis of each site was subjected to public inquiry (in France a formal and routine process without public consultation or participation) during 1997 and, in each case, the Commissioner at the Inquiry pronounced in favour of a laboratory. However, the national assessment board, CNE, the oversight body set up under the 1991 Law, presented its scientific assessment which was favourable to the East and Gard sites but found the Vienne site contained unavoidable negative aspects and complexities compared to other sites (CNE, 1997, p.16). Other evaluations by experts and regulators cast doubt on the Gard site in terms of its constricted size and seismic risks. At the end of 1998 the government recommended in favour of only one site, the East, going ahead. An alternative site (in granite) remained to be identified in Vienne or elsewhere. In the event no granite site was identified although, for a while, the illusion of a comparative assessment was sustained through participation in generic studies of granite formations.

There were already signs of opposition to the proposals. The European Network based in Lyon had sustained a long campaign against the Superphénix. An association of local groups in nuclear communities, the Network relaunched itself in 1997 under the banner 'Sortir du Nucléaire'. It published a survey of the state of the nuclear industry worldwide to demonstrate that France was almost alone in its dedication to nuclear power (*La France isolée dans une Europe sans nucléaire*, May 1997). In 1998 the Network turned its attention to the proposed repository. The action 'Terre vivante' took place at the Bure site with a rally at the nearby town of Neufchâteau. Later in the year there was a demonstration at Verdun. Although these actions were relatively small and were casually dismissed by the nuclear industry they countered the widely held belief that the anti-nuclear movement in France was moribund.

Nevertheless, up to this point the new French approach to site selection had met with qualified success. The 1991 Law was carefully constructed to keep all options open while engaging in a kind of voluntary process of site selection which, in principle, expressed the need for democracy and openness in decision making. It was an attempt to build a consensus recognising the need for social as well as scientific agreement. As Boissac and Tamborini (1999, p.2) observe,

> In modern democracies, certain restrictions can be imposed, where no consensus exists, only after a large debate during which all interests and opinions are allowed to express themselves. That does not mean that scientists and technicians (the so-called 'experts' or 'knowledge holders') must be excluded from the debate, but simply that other arguments than their own must also be heard. Discussions among scientists cannot take only into account rational analyses based on proven facts which exclude hypothetical or non-demonstrable risks, irrational fears and even ethics or moral standards.

Indeed, the conversion to a process that included social and ethical considerations was evident in Andra's publications on risk and on ethics which appeared during the early years of the new century (Andra, 2003, 2004). This more open and socially aware approach became manifest in the development of the Bure site.

Bure – a nuclear no man's land

Bure is a serendipitous location, a place where propitious circumstances appear to coincide. Fundamentally the reason Bure was chosen was geological. Geologically speaking, it sits above a thick layer of argillaceous rock, the Callovo-Oxford clay formation that underlies a wide area in this part of eastern France. The geology of the Paris Basin is well understood and has been stable for a million years. According to Prof. Jean-Claude Duplessy, President of the respected CNE, 'Bure is one of the best sites we might imagine in France' (interview, 2013). Frédéric Cartegnie of Andra explained to me that it lies in a region of oil exploration so it was known that the area had deep, thick, hard clay with a good hydrogeological gradient (interview, 2013). Once the search for a clay site began after the passing of the 1991 law (French Republic, 1991) this area became the leading candidate for the deep research laboratory. The area chosen was determined by specifically geological parameters, gradient, depth, river systems and so on (Andra, 1997a, 2013a).

While geology was the leading consideration in selecting Bure in the Meuse department as a good site for a repository deep underground, above ground other reasons for its selection appear likewise pretty obvious. 'Bure is in the middle of nowhere' (Gerald Ouzounian, interview, 2007), a 'no man's land' in the words of Ben Cramer and Camille Saïsset quoted at the beginning of this chapter (2004). The tiny village of about 80 people together with nearby equally tiny settlements, such as Saudron just over the departmental border in Haute-Marne, is situated in open, rolling countryside with big fields of wheat and barley, some pastureland and swathes of deep green forest. It lies between the wooded and pretty valley of the Marne to the west and the smaller Ornain to the east. But, the overwhelming impression is of space, a deeply rural emptiness with few inhabitants, settlements or traffic. Indeed, the population density here is very low (three people per square kilometre) and absence of people an obvious asset in seeking a site for a nuclear waste repository. 'It's an advantage – not so many people to get to know' says Bernard Faucher, of Andra (interview, 2013). In this hunting country Michel Guéritte, an anti-nuclear activist, claims with realistic exaggeration, 'There are more wild boars than people in this area' (interview, 2013). With just a few small towns nearby (Joinville, St Dizier or Bar-le-Duc) and larger cities such as Nancy more than an hour away, Bure and its surroundings seem isolated and empty, truly reflective of *la France profonde*.

Indeed, Bure is truly peripheral in terms of its remoteness. Although situated within the heart of eastern France, it is an empty heart, a borderland, on the edge

FIGURE 4.12 Rural winter and summer scenes in and around Bure

Source: © ANDRA, insets: author

of geographical, cultural and administrative regions. It is in the internal periphery. Bure lies in the very south of the department of Meuse right on its border with Haute-Marne. It is also a borderland between two regions with Meuse part of Lorraine and Haute-Marne the southernmost part of Champagne-Ardennes.[10] The cultural and economic core of those regions – cities like Reims or Metz or Nancy and the traditional activities like viticulture and metallurgy – which gave them identity are far from Bure.

For long a rural backwater, Bure has, within the space of two decades, begun to take on some of the appurtenances of a modern industrial and commercial park. The headworks and offices of the laboratory, the technological exhibition facility and other buildings (a new hotel and the archives centre of EDF) are now prominent in the lonely landscape. The major part of the development, the underground research laboratory (URL) itself, with its two shafts and network of drifts, is nearly 500m below the surface. In the laboratory, scientific and technical studies are being carried out to determine the containment and retarding properties of the clay, the impact of heat, the feasibility of remote disposition of canisters and containers and the design basis for the repository itself. Under the law, wastes may not be placed in the URL which will not, therefore, be part of the eventual repository but may continue as a large infrastructure for international research (Andra, 2013b). The repository itself, if it is eventually constructed, will be some distance away at a depth of around 500m connected to surface installations by sloping tunnels for transporting machinery, waste and people. Altogether the footprint needed will be 30km² with about half of that required for the laboratory area itself. The whole operation will be remotely handled on the principle, according to Eric Poirot,

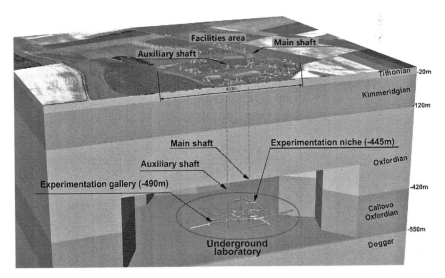

FIGURE 4.13 Schematic diagram of underground laboratory at Bure with surface facilities

Source: © ANDRA

Communications Officer at Bure, that 'the simpler it is, the better it is' (interview, 2013). Although most of the project will be underground, the repository will require extensive surface structures for storing, conditioning and packaging wastes, ventilation shafts, spoil heaps and administrative buildings, together with much improved transport facilities possibly including a reopened railway line. For the first 50 years of its existence (optimistically programmed as 2025–75) the repository will be a 'pilot project' taking wastes initially from Marcoule, France's first reprocessing site in the southern Rhône Valley, before beginning the transfer of the large volumes of wastes in storage at La Hague. Thus, La Hague and Bure are destined to become umbilically connected for generations as the key locations in the long-term management of nuclear wastes through various stages of a continuous (and, in principle at least, reversible for a hundred years) process of managing the country's highly active radioactive wastes. Bure and its region are a peripheral area in the very process of transition from the traditional to the modern.

Haute-Marne and Meuse – a willing and winning combination

Under the 1991 Law (French Republic, 1991), site investigations could not begin without the involvement of 'local officials and members of the public' and the creation and operation of an underground laboratory was contingent on taking account of 'the opinions of the affected municipal, general and regional councils, and following a public hearing' (Articles 6 and 8). Typically, in France, participation in decision making is mediated through the representative political process. In France it may be said that politics is the domain of politicians. While the national government enacts a detailed law setting out the process of staged decision making, as in the case of the formative Bataille Law of 1991, the law is administered through the local political machinery of the departments and the communes. The wider public may contribute and possibly achieve influence through occasional consultative procedures such as the national debates (two of which, in 2005–6 and 2013, have been on radioactive waste), local forums and hearings. In practice, decision making is effectively confined to elected or appointed representatives. As Benoît Jaquet, Secretary-General of the local information and monitoring committee (CLIS), put it, when we first met in 2005, 'We are following an old point of view, a centralised state. It is the French system', representative but not particularly participative. In this tradition the mediation mission under Christian Bataille had undertaken the search for sites which eventually led him to select the neighbouring departments of Haute-Marne and Meuse in eastern France. At this point the departments became the prime movers in the site selection process.

At first Haute-Marne was the most anxious to proceed. According to Bernard Faucher of Andra who was involved in community liaison during the early years of Bure, Haute-Marne regretted that it had missed out on all the tax and other benefits that had flowed into the neighbouring Aube department which had hosted the national low level waste repository, Centre de l'Aube at Soulaines. In

Haute-Marne socio-economic reasons were the motivation; the department had low employment and was a veritable desert in terms of higher education. Five counties in northern Haute-Marne were identified as potentially suitable by the French Geological Survey and both the department and the local mayors were in favour. While Haute-Marne made the early running, in Meuse there was no debate and no response to the call for volunteers until it felt it might lose out. Faucher claims that ANDRA created rivalry between the two departments. Meanwhile, the geological surveys indicated land within Meuse to be most suitable for the laboratory though surface facilities could be in both departments. The potential rivalry was solved by the political expedient of merging the relevant parts of the two departments into a single research area henceforth called the East site or more popularly the Bure site.

An area at the margin

The Bure site was achieved relatively easily. The geology was highly promising and in this empty, underdeveloped region the prospect of an underground laboratory was widely welcomed at the political level for the economic benefits it might bring. Indeed, economic benefits are an integral part of the package that goes with major infrastructural developments in France and, in the case of the laboratory, were enshrined in the law. Under Article 12 of the 1991 Law, a Public Interest Group (Groupement d'Intérêt Publique or GIP) could be formed to manage resources, effectively to distribute public funding from waste producers split equally between the two departments. Regarded as compensation by government and ANDRA (interviews with Marc-Antoine Martin, 2005, Gerald Ouzounian, 2007) in the sense of investment to enable communities to host a facility and to contribute to well-being, the funds are deployed to support local infrastructure, social housing, economic development, environmental projects and tourism. The GIP resources amount to a substantial and continuing injection to areas within 10km of the site which receive as much as €400 per inhabitant per year. Eric Chagneau, a former Director of the Fund, described it as 'a national programme in a small part of France' where the capacity for development was low. 'We considered it would be dangerous to put all the money around Bure. You have a little engine and if you put too much gas in it, it can't go any faster' (interview, 2005). This problem of underdevelopment was echoed by Bernard Féry, coordinator of a programme for major project investment associated with the development of the laboratory (interview, 2005). For him, in the early years, it was hard going trying to get projects such as a scientific research centre off the ground. 'Development here is very, very, very difficult,' he emphasised. It was difficult to entice people to visit, there was a lack of know how in a backward area that was 'très, très rurale'. He had concluded it was an economic backwater without possibilities. The economic marginality of Bure was profound indeed.

Nevertheless, financial inducements and the promise of prosperity were clearly fundamental to the relatively quiescent emergence of the repository project in this rural outback. Michel Guéritte who campaigned against the low level radioactive waste Centre de Stockage at Soulaines recognised the pull of money on a small community such as Soulaines in neighbouring Aube department. So, too, in Bure 'there is no problem. We have the money. The district is wooed with the money of ANDRA' (interview, 2013). There was virtual universal, though sometimes wary, welcome among the political elites at regional, departmental and communal level. A general ambivalence towards the laboratory among the community was summed up for me by M. Robert Fernbach, the Mayor of Houdelaincourt (Meuse) back in 2005:

> We need to let research continue although there is concern about only one laboratory here in Bure, rather than another for comparison. People are on the whole indifferent as the economic problems mean they have other preoccupations. The nuclear industry is associated with the nuclear dustbin but people want to live here, to have work and to have activities. The project could bring development and people will have to find new ways.

In this respect the idea of 'poles of excellence' with linkages to Lorraine and an education centre based on the laboratory at Bure keep alive the fragile vision of a broader economic dimension of which Bure is a potential epicentre.

Fatalistic acceptance

The acceptability of Bure in the local community was not just simply a case of 'The invader who comes with money', in the words of Claire Mays (interview, 2004). Rather, ANDRA came more like a thief in the night, by stealth and slow degrees. Its arrival was likened to 'a type of Trojan Horse' by Frédéric Merillier of Greenpeace (2004). The establishment of the laboratory was achieved in a measured, deliberate fashion as, by subtle changes, the project has moved from being one of possibly several sites, to a single site for a research laboratory, to its projected status as the nation's deep geological repository for radioactive wastes – Le Projet Cigéo. One by one significant changes have been made which have narrowed the options to the one single objective, moving from the possible to the inevitable. And, during the process, the public reaction (in so far as it can be gauged or generalised) has also moved, from ignorance to fatalistic acceptance, to the point where, 'They know it's going to happen and they can't do much about it' (Faucher, interview, 2013). Bure is already showing the manifestations of the kind of social acceptance that are a characteristic of peripheral locations as we have seen in the previous chapters. A major reason for this acceptance of the inevitable has been the way that the notion of geological disposal of high and medium level wastes at the Bure site has been progressively embedded in the community as other options have, one by one, disappeared. It is possible to identify five aspects of this narrowing of options.

Firstly, the geographical options have narrowed to one site only. Eastern France was, under the 1991 Law, merely one of the options for reversible or irreversible disposal in deep geological formations. The East of France URL was identified as the clay site to be compared with the granite site in Vienne near Poitiers in west-central France (ANDRA, 1997b) and the site in a silt formation in Gard in the Rhône Valley near the reprocessing works at Marcoule (ANDRA, 1997c). By 1998, as we have seen, both these sites had been dropped. The CNE was instrumental in rejecting the Vienne site on the grounds of fast water circulation and the problem of finding an acceptable block of granite. Prof. Jean-Claude Duplessy, President of the CNE, described the Poitou site to me as, 'The worst I could imagine. But, people were upset by the decision. It was strongly supported locally' (interview, 2013). The southern site was also abandoned though, according to Duplessy, 'it was not a bad site but wine-growers were concerned and government forgot about it'. So, the other options disappeared, one for geological, the other mainly for social reasons. For a time the idea of finding a granite site for comparison ('just to keep people happy', Duplessy) was kept alive though it segued into a generic rather than site-specific concept and it, too, was dropped. So, by the time of the new Law in 2006, the East site was the sole site for continuing research to design and implement a future repository (ANDRA, 2006).

Secondly, there has been an increasing focus on geological disposal as the only method of long-term management. Of the three axes of research defined under the 1991 Law, the first, partitioning and transmutation (P and T), continues to be pursued but in the context of possibilities of implementation in conjunction with R and D on a future (fourth, fast breeder) generation of reactors (CNE, 2012). So, while progress has been made to determine that partitioning is working and transmutation can be envisaged, P and T is still very much work in progress in the realm of theoretical work with potential future application rather than a commitment to industrial scale development. In any event P and T offers only a partial solution to the waste problem reducing but not eliminating the volumes and radioactivity. The third axis, long-term storage, is not considered to be an option in its own right, rather a complementary stage in an integrated process leading ultimately to permanent disposal. Consequently, it is the second axis, evaluation of options for deep disposal, that has become, *de facto*, the primary, indeed, the only method for long-term management of highly active nuclear wastes. The CNE has given its imprimatur to geological disposal as 'absolutely necessary' (Duplessy, interview, 2013) and the 2006 Law spelled out the milestones with a site licensed by 2015 and a repository commissioned in 2025. However, it has become increasingly clear that disposal will also be combined with storage on site involving the need for considerable space for surface stores for wastes prior to disposal and for conditioning, packaging and encapsulation facilities necessary to prepare wastes for permanent emplacement.

Thirdly, in the effort to make this method more socially acceptable, reversibility was introduced as a requirement of deep disposal. The concept and implementation of reversibility and retrievability have been subjects of considerable debate in France (CNE, 1998; Aparicio, 2010). Retrievable disposal was confirmed in the 1991 Law (Article 2) though unlimited and, therefore, non-retrievable disposal was not outlawed (Article 4). The 2006 Law went quite explicitly for 'reversible waste disposal in a deep geological formation' (Article 3). Under the law the concept of reversibility was to be built in to indicate that, though geological disposal was the preferred option, it was not necessarily the only option, that is, until the repository was finally closed. It is a kind of staged process. The problem with reversibility as a way of keeping options open is that the space for exercising it gradually narrows. As sociologist Yannick Barthe put it to me during our conversation in 2013, 'Reversibility suggests that you can go back – but reversibility is also less and less reversible'. But it may contribute to social acceptability. Barthe went on, 'A more democratic decision is a more reversible decision. Democracy is about choice and ability to change. Industry does not like reversibility – it is a compromise. Reversibility is open and avoids a decision that is irreversible'. On the other hand, it is not clear when, if ever, the time would be ripe for that irreversible decision (final closure) to be taken (Barthe, 2010). The point about reversibility, as Bernard Faucher notes, is that 'You need political reversibility. We want to be able to say No' (interview, 2013).

Fourthly, the status of the project gradually has changed over the years (ANDRA, 2013b; ANDRA, undated). Initially, under the Bataille mission, sites, including east France, were identified and then, from 1994, geological investigation by ANDRA commenced. By the turn of the century construction of the URL had begun and reversibility was under consideration (CNE, 1998). By the time of the 2006 Law it was clear the design would include reversibility and in 2014 ANDRA confirmed a 'phased approach' allowing the next generation to decide how to proceed (ANDRA, 2014). The URL would remain a research facility with a repository located separately although nearby. The repository itself from its date of commissioning for at least 50 years would, in Frédéric Cartegnie of ANDRA's words, be a 'pilot project', maintaining the idea of temporariness while the reality was that of permanence. As with reversibility, so with disposal, the French approach is softly, softly as key decisions are taken and options foreclosed.

Fifthly, the role of the repository in terms of its potential inventory has gradually expanded thus reinforcing its position as the only option available for long-term waste management. Originally, the repository was intended for the high active (HA) wastes mainly (84%) from the electricity generating programme but also some reprocessed waste arising from defence and research programmes. These wastes are converted into glass blocks and mainly stored at La Hague representing 0.2% of the volume but 96% of the radioactivity of all wastes. To these were added wastes of medium activity but long life (MA-VL) also emanating from reprocessing covering 3% of the volume and 4% of the radioactivity. Most of the remainder, lower activity, short-lived wastes, more than 90% of the total volume, goes into the near surface repositories (Centre de l'Aube for low level and nearby Morvilliers for very low level wastes) in the nearby department of Aube (ANDRA, 2013a). During the pilot phase of Cigéo it is intended that the repository will take medium level wastes, then high level wastes from Marcoule before receiving high level wastes from La Hague after they have cooled sufficiently from around 2075. But, there is uncertainty over the future inventory. Much depends on future nuclear policy. As was indicated earlier it is not clear whether unreprocessed spent fuel may eventually be destined for direct disposal, what may be done with surplus plutonium stocks, what is the eventual strategy for MOX fuel should it become a waste rather than a resource. The truth is that no one knows what the eventual inventory will be, adding yet another uncertainty to the future of the project.

Geological disposal has become the main option, clay the favoured formation and Bure the single site. The project has moved from a research laboratory towards an industrial pilot phase and ultimately towards a fully fledged repository. Although reversibility has been introduced as a means of keeping options open, the scope for reversal will decline. There is, in reality, according to Barthe, a process of 'lock in' despite the rhetoric that suggests otherwise. There is an 'escalation of commitment' whereby a process is controlled and gradually reduces the range of possibilities (Barthe, 2010). Meanwhile, the volumes, types and characteristics of the inventory are uncertain. This means that the repository concept will have to be adapted and modified as the composition of the inventory becomes clearer and, if spent fuel is reclassified as a waste with a consequent increase in the volume of the inventory, the changes could be considerable perhaps requiring the creation of a new facility.

Outright opposition to the Bure project has been fitful and most in evidence in the early years. At the site itself opposition has been conducted in specific manifestations organised from the towns and cities like Bar-le-Duc or Nancy further away or expressed through national groups or the national public debates that are organised from time to time. The protest group Burestop has held a number of actions and events and local groups affiliated to Sortir du Nucléaire have been locally active from time to time. But, given the lack of local population, activity near the site is difficult to mobilise on a continuing basis. Over time, in the two affected departments, most opposition has become institutionalised in the Clis, the Comité Local d'Information et de Suivi (local information and oversight committee).

The Clis was established under Article 14 of the 1991 Act. It has a membership of 90 comprising local representatives of government (MPs, Senate), region (Lorraine and Champagne-Ardennes), departments (Meuse and Haute-Marne), communes (mayors), Chambers of Commerce, trade unions, agricultural interests and environmental groups and is chaired by a departmental representative. Initially conceived as a consultative body responding to plans and proposals for the laboratory, the Clis became frustrated with its role. According to its Secretary-General, Benoît Jaquet, it developed a 'role neither foreseen nor desired. The Clis doesn't have a place in the decision making process – so it must make its place' (interview, 2005). On the face of it, the Clis has been pretty energetic over the years, holding regular meetings in public, compiling and disseminating information, responding to consultations, undertaking investigations, providing critical assessments of proposals and so on. Among its activities has been promoting the idea of reversibility and commissioning a report from a group of international experts led by Dr Arjun Makhijani of the Institute for Energy and Environmental Research in Washington DC. This was a critical evaluation which expressed reservations on geological grounds that the Bure site could be pronounced with certainty as adequate to proceed to a decision under the law in 2006 as a site for a possible geological repository (IEER, 2004; Clis, 2005). In a subsequent critical review in 2011 the IEER indicated its most serious concern to be 'a pervasive optimism in the interpretation of complex phenomena with regard to repository performance' and it again voiced a concern that the timetable was far too rushed given the amount of research and characterisation that remained to be done (IEER, 2011, p.12).

The extent of influence of the Clis is difficult to gauge except in general terms. It certainly raises public awareness but mainly among the political and stakeholder class rather than the general public. 'Clis is composed of many interests and so it is difficult to get one point of view' (André Mourat, member of Clis, 2005). Some, especially from the perspective of the nuclear industry, regard Clis as a delaying mechanism. The role of the Clis is to assemble the local actors in a consultative assembly. From his perspective as Chairman of the Clis, Jean-Louis Canova considered its role to be to understand concerns and to gather information both from ANDRA and independent experts and to bring ANDRA to account (interview, 2013). It may well be that there is general confidence in experts and the Clis. But, there is a general expectation that Cigéo is here to stay and the debates are all part of the theatre of decision making. Despite its critical approach the Clis is circumscribed in power and thereby helps to perform a legitimating function for the project.

Among the broader population in the Bure area and beyond there appears to be a culture of resignation, acceptance and, perhaps too, a sense of patriotism and pride in sacrifice. This is eastern France, the region of warfare and resistance. As Bernard Faucher expressed the sentiment, 'We had Verdun, we had Sedan, we are tough people – see what we are ready to do for France'. Michel Guéritte draws a dismal parallel between war and waste. 'This area is the nuclear waste region of France. At Verdun there were a million dead. Bure could have a similar impact. Bure is a monument to death.' To a crowd at Bure on Armistice Day he proclaimed, 'Soon, Bure will be more famous than Verdun'. But, for the present and for the most part, Bure remains silent, rather untroubled and hidden far away from the mainstream but a place, nonetheless, which like La Hague is shaping the nuclear destiny of France.

Communities on the edge and at the centre

La Hague and Bure are two places that I have found compelling both in their geography and recent history. They are places at once on the edge and also at the centre. La Hague is on the furthest edge of the country; Bure is in an empty borderland in the depths of rural France. Yet, both are also at the centre, they are at the very heart of France's nuclear project: the one, La Hague, the fulcrum around which the closed French nuclear cycle revolves; the other, Bure, the place where, in future years, the dangerous and useless output of the industry is likely to be laid to rest. They are places that reflect and reproduce the transformations that take place as a modern industrial activity is introduced into a traditional community. In the northern Cotentin where La Hague is situated, the process of modernisation wrought by the arrival of the reprocessing plants has reached its maturity. Initial resistance and ambivalence towards the new has eventually led through gradual absorption and acceptance to a new tradition, as it were, based on a combination of coexistence and integration. Traditional integrating processes of family, community, trade unions, work and voluntary organisations that previously existed within the traditional societies are also developing as modernising processes mature and create or recreate processes of integration within the wider society.

Bure is at a much earlier point, at the very cusp of the transformative process where a 'backward' area is confronted with the almost alien intrusion of the modern. The nuclear industry is physically visible though the major part of the activity is underground but there is little evidence of social impact in the vicinity. It is the very silence of its presence in a remote district that has enabled the laboratory to gain its foothold. Its imprint is, for the present, both light and dispersed around the surrounding region. In the years ahead as activity intensifies there will be more evidence of infrastructural development around the site itself but the social impacts are likely to be gradual and dispersed throughout a wide area thereby facilitating a barely perceptible process of integration. In both these communities the process of peripheralisation may be invoked to both describe and explain the transformations that occur and the implications for policy development.

At La Hague the conflicts of early years have disappeared and have been replaced by quiescence as the industry has achieved economic pre-eminence and achieved a social transformation of the area. La Hague is the archetypal expression of the peripheral community – remote, dependent, relatively powerless and socially perhaps somewhat introspective. Coming into a vulnerable community the nuclear industry was able to develop the world's largest civil reprocessing complex. Although there was conflict, on the whole it was not directed at the reprocessing itself but rather at its function as the centre of an international trade in nuclear materials. In La Hague conditions of remoteness, economic dependence and a community at once both powerless and defensive instil a resilience necessary for its survival. In this setting the traditional (farming and fishing) and the modern (industry) do not simply exist separately side by side, they coexist with sufficient integration to suggest that the process of modernisation, far from overwhelming traditional institutions, has, in some senses, reinforced them. The sense of collective identity expressed in feelings of community or organisation in trade unions indicates a solidarity that can withstand external threats. The industry may only have been accepted with reluctance in parts of the community but it has been accommodated.

Bure, too, provides an optimal setting for the management of the nuclear industry's most unwanted product, its waste. Here, the process of peripheralisation is being actively if unconsciously developed by the state and the nuclear industry. A remote site has been selected, economic incentives and development have been put in train, the local political class has, on the whole, been suborned and incorporated, opposition has been marginalised or co-opted and environmental and health risks have been played down. At Bure we are witnessing the essence of an approach to policy making which recognised the necessity of local acceptability. Bure reflects the awareness that the long-term management of radioactive waste depends on the willing compliance of the communities that host it. In a highly centralised country the political necessity of the interdependence of the local and the national is confirmed. For just as Bure, and other nuclear communities are created and sustained by policy enacted by the state, so, too, the central state is dependent on the peripheral communities for the successful implementation of its policies. For the communities hosting a repository the local employment directly created will be relatively modest, highest during the construction phase but falling once the repository is complete. The adventitious population created by the investment is likely to be small though the economic impacts in the surrounding region will be substantial. But, in return for the economic and social benefits the peripheral community gains, it yields a territory which legitimises the continuation of nuclear development and bears a burden that will be present for generations long after the industry has disappeared.

The two communities reflect the necessity of a local and social dimension to radioactive waste policy making. La Hague has experienced the discursive shifts in nuclear power that occurred during the last decades of the twentieth century. In the early years in France, as elsewhere, the discourse of Trust in Technology was dominant and a strategy, if such it can be called, of Decide Announce Defend prevailed in the siting of nuclear facilities. La Hague was the product of this period when the site for the reprocessing works was identified and developed in relative secrecy. In the decades that followed the appearance of the plants in the Cotentin, anti-nuclear protests became confrontational and focused, especially on Flamanville and on the trade in nuclear materials as La Hague became an international target during the discourse of Danger and Distrust. As the discourse of Cooperation and Consensus evolved during the 1990s and into the new century, La Hague was becoming more embedded in the local community and controversies over reprocessing had more national than local resonance.

This was the period of Bure's gestation, a period when France, like other nuclear countries in the West, was experimenting with the new commitments to openness and dialogue that became characteristic of nuclear policy making during the deliberative turn. In fact, in France it is possible to date this turn rather precisely. After the failure of the Castaing Commission to find a disposal site for high level wastes, the Bataille Law was enacted in the 1990s ushering in a new era of policy making based on the idea that social consent was as necessary as scientific confidence in finding a place to put waste. The subsequent Bataille Mission's stated objectives were: to circulate information; to facilitate decisions; and to open dialogue. 'The Mission responds to all offers of dialogue, listens to all concerns, and goes to each interested Department to hear expectations, questions and comments' (Bataille, 1993, p.6). The 1991 Law was carefully conceived to provide adequate time for resolution with research on three alternatives over a span of 15 years before conclusions were reached and an option chosen. The need for equity was recognised in the various proposals for consultation and compensation. In these various ways it was intended that a sufficient consensus, both locally and more widely, would be developed to enable an agreement to proceed.

Of course, it was not so straightforward and it may be said the French interpretation of the deliberative turn was more circumspect and restrained than was the case in contemporary Sweden or the UK. The national debates in France suggest a commitment to participation at the national level that is stronger than elsewhere. But, at the local level, the French approach to voluntarism relies heavily on the representative democratic process and the deployment of resources and commitment to economic development quite widely dispersed to ensure support. In contrast, participative democracy at the local level is much weaker than in some other countries and processes of engagement with citizens and other stakeholders are almost entirely lacking. Bure also exemplifies a somewhat pragmatic approach to policy making whereby a voluntary process is, in reality, more *dirigiste* as the options are narrowed and the site becomes locked in to its inevitable role. Despite the intention of the law, Bure was not subject to a comparative assessment but became the only site under serious consideration. Although it was only a 'laboratory' and in its early operational phase it will be a 'pilot' facility, there is little doubt that Bure will become a permanent repository. There are further uncertainties that lie ahead, notably what the ultimate inventory of wastes will be but, as it stands it is highly likely that Bure will fulfil its putative role as the place for the permanent burial of the nation's more dangerous radioactive wastes. The French approach to locating its nuclear facilities is, on the one hand, teleological and unswerving but, on the other, shows an appreciation for responsiveness and accommodation that ensures successful implementation in the long run. In seeking to integrate the technical and the political aspects of decision making the French have been seeking 'the way to decide without making a definitive decision' (Barthe, 2009, p.952). Although nothing in the management of radioactive wastes can ever be certain, the policy inertia that maintains reprocessing at La Hague seems also likely to underlie the continuing commitment to Bure as the place for the final solution of the problem of radioactive waste.

Visits and interviews

Throughout this chapter I have recorded the views of many people I met during several visits to La Hague, Bure and Paris. Below are details of some of the visits and people I met who helped provide the ideas and substance of this chapter. I am grateful to all of them.

1986 La Hague, County Councils Coalition

The visit was part of a fact finding tour of Sweden, Germany and France organised by the County Councils Coalition (Bedfordshire, Lincolnshire and Humberside) in

October 1986 to gain comparative information on facilities and approaches to radioactive waste management. The findings are recorded in *The Disposal of Radioactive Waste in Sweden, West Germany and France* (Environmental Resources Ltd., January, 1987).

1993 Centre de l'Aube, Marcoule, Paris, La Hague. Radioactive Waste Management Advisory Committee (RWMAC)

A study tour organised by RWMAC in September 1993 included visits to Centre de l'Aube, the low level near surface disposal facility in eastern France, the reprocessing plants at Marcoule in the Rhône Valley in southern France and at La Hague in Normandy. A report on the tour, *Study Tour of France,* was published by RWMAC in June 1994.

1996 visit to Paris and La Hague

Interviewed in Paris May, 1996: M. Barthelemy, Ministry of Transport; Robin des Bois, Environmental NGO; Mycle Schneider, World Information Service on Energy; Jean-Luc Thierry, Greenpeace France.

Interviewed in Caen: Pierre Barbey and David Boilley, Association pour le Contrôle de la Radioactivité dans l'Ouest (ACRO).

Interviewed in Cherbourg: Yves Bonnet, Deputy for La Manche; Ghislain Quetel and André Guillemette, trade union representatives (CFDT); Didier Anger, Comité de Reflexion d'Information et de Lutte Anti Nucléaire (CRILAN); David Bosquet, Green Party.

Interviewed at La Hague: Patrick Fauchon, COGEMA La Hague.

1999 visit to Paris and La Hague

Interviewed in Paris, May 1999: Keith Shannon, British Embassy; Philippe Pradel, COGEMA; Jacques Tamborini, ANDRA; Arsène Saas, Commission Nationale D'Evaluation (CNE).

Interviewed in La Hague: Philippe Fournier and Jean-Guy Devezeaux de Lavergne, COGEMA.

Interviewed in Cherbourg: Yannick Rousselet, Greenpeace.

2004 visit to Paris, CoRWM

Interviewed: Yves le Bas, ANDRA; Yves Marignac, WISE; Claire Mays, consultant; Ben Cramer, journalist; Hugh Elliott, British Embassy; Frédéric Merillier, Greenpeace; Jean-Marc Péres, Institut de Radioprotection et de Sûreté Nucléaire (IRSN); Jean-Pierre Chaussade, Débat Publique, EdF; Prof. R. Guillaumont, Jean-Paul Schapira and Rémi Portal, CNE.

March 2005, visit to Bure

Interviewed: Marc-Antoine Martin, ANDRA Bure Laboratory; Eric Chagneau, Groupement d'Interêt Publique (GIP); Bernard Féry, coordinator of economic development; Benoît Jaquet, Secretary-General, Comité Locale et de Suivi (CLS) at Bure; Sylvie Malfait-Benni, Robert Fernbach, Jean Coudry, Jean-Marie Malimgreau, André Mourot and Sandrine Soehnlen, members of the CLIS.

May 2005, visit to La Hague

Interviewed in Cherbourg, May 2005: Didier Anger.
 Interviewed in La Hague: Jean-Michel Maghe, Program Manager, COGEMA.

February 2007, visit to Paris

Interviewed: Claire Mays, consultant; Gerald Ouzounian, ANDRA.

October 2013, visit to Bure, Paris and Cherbourg

Interviewed in Bure, October 2013: Frédéric Cartegnie, ANDRA; Jean-Louis Canova and Benoît Jaquet, Clis; Eric Poirot, ANDRA; Michel Guéritte; Patricia Andriot, Conseil Général Champagne-Ardennes.
 Interviewed in Paris: Jean-Claude Duplessy, CNE; Yannick Barthe, Centre for Sociology of Innovation; Yves Marignac, WISE-Paris; Bernard Faucher, ANDRA; Luis Aparicio, ANDRA; Claire Mays, Consultant.
 Interviewed in Cherbourg: André Guillemette, ACRO; Didier Anger, CRILAN.

Notes

1 The reprocessing plant at Marcoule in the Rhône Valley produced plutonium for military purposes and was closed in 1997.
2 Areva merged with COGEMA (Compagnie Génerale des Matières Nucléaires) in 2001 taking on the former company's role in providing fuel enrichment and production, reprocessing facilities and waste storage and conditioning facilities.
3 Agence nationale pour la gestion des déchets radioactifs (French national radioactive waste management agency). ANDRA is an independent public agency with funding drawn from nuclear operators.
4 Comité de Réflexion, d'Information et de Lutte Anti-Nucléaire.
5 Association pour le Contrôle de la Radioactivité dans l'Ouest.
6 Confédération Française Démocratique du Travail.
7 Research in Radioactive Waste Management, Law No. 91-1381 of December 30, 1991, Article 3. See reference, French Republic, 1991.
8 'The Cotentin is like the ostrich. It puts its head in the sand. It doesn't see the hunter, but the hunter fires into its backside with his gun.'
9 Commission Nationale D'Evaluation Relative aux Recherches sur la Gestion des Déchets Radioactifs.

10 The two regions together with Alsace became part of a new super administrative region under territorial reforms introduced in 2016. Bure is on the border between the two traditional regions of Champagne and Lorraine.

References

ANDRA (Agence Nationale pour la Gestion des Déchets Radioactifs) (1997a) *Research Laboratory, State of Knowledge and Experimental Programme, East of France*, Andra, Châtenay-Malabry.

ANDRA (1997b) *Research Laboratory, State of Knowledge and Experimental Programme, Vienne*, Andra, Châtenay-Malabry.

ANDRA (1997c) *Research Laboratory, State of Knowledge and Experimental Programme, Gard*, Andra, Châtenay-Malabry.

ANDRA (2003) *Faut-il avoir peur des déchets radioactifs?* Andra, Châtenay-Malabry.

ANDRA (2004) *Ya-t-il une ethique de la gestion des déchets radioactifs?* Andra, Vuibert, Château-Malabry and Paris.

ANDRA (2006) *Radioactive Materials and Waste*, Planning Act of 28 June 2006, Act No. 2006-739, Consolidated version established by Andra.

ANDRA (2013a) Cigéo: l'Histoire d'un Projet, Cigéomag, Supplement thématique au Journal de l'ANDRA, August.

ANDRA (2013b) *Project Cigéo Centre Industriel de Stockage Réversible Profond de Déchets Radioactifs en Meuse/Haute-Marne*, Andra, Châtenay-Malabry.

ANDRA (2014) 'Andra presents the actions to be taken following the public debate on the Cigéo project', Press Release, 16 May.

ANDRA (undated) *The Presence of Andra in the Meuse and Haute-Marne Districts*, Meuse-Haut-Marne Centre, Bure, France.

Anger, D. (1987) *Silence, on Contamine*, Les Pieux, France, Claude Turpin.

Aparicio, L. (ed.) (2010) *Making Nuclear Waste Governable: Deep Underground Disposal and the Challenge of Reversibility*, Paris, Springer and Andra.

Barthe, Y. (2009) 'Framing nuclear waste as a political issue in France', *Journal of Risk Research*, 12, 7–8, October–December, 941–954.

Barthe, Y. (2010) 'Nuclear waste: the meaning of decision-making', in Aparicio, L. (ed.) (2010) *Making Nuclear Waste Governable: Deep Underground Disposal and the Challenge of Reversibility*, Paris, Springer and Andra, pp.9–27.

Bastard, G. Le and McMurphy, M. (1995) 'Optimum management of weapons plutonium through MOX recycling', *ASME 5th International Conference on Radioactive Waste Management and Environmental Remediation*, Berlin, 3–9 September.

Bataille, C, (1990) *Report on the Management of Highly Active Nuclear Wastes*, Parliamentary Office for the Evaluation of Scientific and Technological Choices, National Assembly No. 1839, Senate, 184.

Bataille, C. (1993) *Mission Report*, Mediation Mission on Siting Underground Research Laboratories, Paris, December 20.

Blowers, A. and Leroy, P. (1994) 'Power, politics and environmental inequality: a theoretical analysis of the process of "peripheralisation"', *Environmental Politics*, 3, 2, Summer, 197–228.

Blowers, A. and Lowry, D. (1996) 'Nuclear transportation – the global vision', *Proceedings of the International Topical Meeting on Nuclear and Hazardous Waste Management*, Seattle, American Nuclear Society.

Boissac, E. and Tamborini, J. (1999) *Proceedings of WM '99*, WM Symposia Inc., Tucson, Arizona, 28 February to 4 March, 45/5.

Boyle, M. and Robinson, M. (1987) 'Nuclear energy in France: a foretaste of the future?', in Blowers, A. and Pepper, D. (eds) *Nuclear Power in Crisis,* London, Croom Helm, pp.55–84.

BBC (2011) 'Nuclear power gets little support world wide', BBC News Science and Environment 25, November.

Cézanne-Bert, P. and Chateaureynaud, F. (2010) 'The argumentative trajectory of reversibility in radioactive waste management', in Aparicio, L. (ed.), *Making Nuclear Waste Governable*, Paris, Andra, Châtenay-Malabry.

CNE (Commission Nationale d'Evaluation, National Assessment Board) (1997) Asssessment Report No. 3, instituted by the law 91-1381 of 30 December 1991, Paris, September.

CNE (Commission Nationale d'Evaluation, National Assessment Board) (1998) *Reflexions sur la reversibilité des stockages*, Paris, June.

CNE (2012) Assessment Report No. 6, instituted by the law no. 2006-739 of 28 June 2006, Paris, November.

Clis (Comité Local d'Information et de Suivi) (2005) *La Lettre du Clis*, Clis, 40 rue du Bourg, Bar-le-Duc, July.

CoRWM (2008) *The Overseas Experience of Radioactive Waste Management*, CoRWM Document Number 2213.1, February.

Compagnie Générale des Matières Nucléaires (COGEMA) (1997) *Annual Report,* Velizy-Villacoublay, France.

County Councils Coalition (1987) *The Disposal of Radioactive Waste in Sweden, West Germany and France,* prepared for the County Councils Coalition by Environmental Resources Ltd.

Cramer, B. and Saïsset, C. (2004) *La Descente aux Enfers Nucléaires,* L'Esprit Frappeur, Paris.

Davis, M. (2002) *La France Nucléaire, matières et sites*, Paris, Wise-Paris.

Environment Agency (2010) Generic design assessment UK EPR nuclear power plant design by Areva NP SS and Electricité de France SA, Bristol, June.

Eurobarometer (2005) *Radioactive Waste*, Special Eurobarometer 227/Wave 63.2, European Commisssion Brussels, September.

Eurobarometer (2010) *Europeans and Nuclear Safety*, Report, Special Eurobarometer 324, European Commission, Brusssels.

Foratom (2014) *What People Really Think About Nuclear Energy*, report, European Atomic Forum, September, Brussels.

French Republic (1991) Official Journal, Laws and Decrees, Law No. 91-1381 of December 30, 1991 on Radioactive Waste Management Research.

Giraud, J. and de L'Epine, Ph. (1995) 'Spent fuel management: present status and future trends of reprocessing and recycling', *Fifth International Conference on Radioactive Waste Management and Environmental Remediation*, Berlin, 3–7 September.

Guillemette, M. (2012) *Iode 129, rejets associés aux opérations de retraitement des combustibles irradiés et données environmentales*, October.

Guillemette, M. and Zerbib, J-C. (2012) *La Hague, territoire sous influence des rejets de carbon 14 des usines de retraitement de La Hague*, Étude de l'incidence des rejets de carbone 14 sur l'environnement terrestre proche (1 à 10 km) et sur l'environnement marin du site Areva NC de La Hague, October.

Hecht, G. (1998) *The Radiance of France: Nuclear Power and National Identity after World War II,* Cambridge Mass. and London, The MIT Press.

HSE (Health and Safety Executive) (2009) *Public Report on the Generic Design Assessment of New Nuclear Reactor Designs*, Areva NP SAS and Electricité de France SA EPR Nuclear Reactor, HSE, Bootle.

Hochman, N. and Decoin, D. (1997) *A Photographic Journey in La Hague*, Cherbourg, Editions Isoète.

Homberg, F., Pavageau, M. and Schneider, M. (1994) *Cogema-La Hague: The Waste Production Techniques*, World Information Service on Energy (WISE), Paris, December.

IEER (Institute for Energy and Environmental Research) (2004) *Examen critique du programme de recherche de l'ANDRA pour déterminer l'aptitude du site de Bure au confinement géologique des déchets à haute activité et à vie longue*, Rapport Final, IEER, Takoma Park, Maryland, USA, December.

IEER (2011) *Critical Review of Andra's Program of Research Conducted in the Underground Laboratory at Bure and in the Transposition Zone to Define the ZIRA*, Final Report, IEER, Takoma Park, Maryland, USA, March.

IFOP (2011) *Les Français et le nucléaire*, poll commissioned by *Le Journal du Dimanche*, June.

IPFM (International Panel on Fissile Materials) (2011) *Managing Spent Fuel from Nuclear Power Reactors, Experience and Lessons from Around the World*, IPFM, September.

Mays, C. (2004) 'Where doe it go: siting methods and social representations of radioactive waste management in France', in Boholm, A. and Lofstedt, R., *Facility Siting: Risk, Power and Identity in Land-use Planning*, London, Earthscan.

Mays, C. and Poumadère, M. (1996) 'Uncertain communication: institutional discourse in nuclear wate repository siting', in Sublet, G., Covello, V, and Tinker, T. (eds), *Scientific Uncertainty and Its Influence on the Public Communications Process*, Dordrecht, Kluwer NATO Advanced Research Series.

Radioactive Waste Management Advisory Committee (RWMAC) (1994) *A Study Tour of France*, London, June.

Ridley, F. (2006) 'French technocracy and comparative government', *Political Studies*, 14, 1, 34–52, February.

Schneider, M. (2013) 'France's great energy debate', *Bulletin of the Atomic Scientists*, January/ February, 69, 27–35.

Sortir du Nucléaire (1998) *Hors-série*, 3, June.

Sovacool, B. and Valentine, S. (2012) *The National Politics of Nuclear Power, Economics, Security and Governance*, London, Routledge.

The Typewriter (2013) 'Is France ready to abandon its nuclear plants?', *Online Category Europe*, 16 December.

WISE (2015) *The French Nuclear Industry in Deadlock: The Burden of France's Nuclear Gamble in the Era of the Energy Transition*, WISE-Paris, 23 June.

WNN (World Nuclear News) (2013) 'Growing support for nuclear France', 24 June.

Zonabend, F. (1993) *The Nuclear Peninsula*, Cambridge, Cambridge University Press.

5

GORLEBEN, GERMANY

The power of resistance

> For the Wends were mostly peasants, like the Scandinavians; tillers and herdsmen living in small villages and raising corn, flax, poultry and cattle. With fishing, bee-keeping and trapping as side-lines.
>
> *(Eric Christiansen,* The Northern Crusades*, 1998, pp.27–8)*

Introduction

Close by the River Elbe, about 130km upstream from Hamburg in the region known as the Wendland, is the small village of Gorleben. It appears an unremarkable, tranquil north German village of sturdy houses, well-paved, clean streets with a few shops and a sports complex. It is surrounded by the well-kept arable farmland and the forest and heath which are typical of this part of the north German Plain. Yet, the name Gorleben is synonymous with the most vigorous, sustained and unyielding anti-nuclear protest experienced anywhere. This obscure region has been at the very heart of the conflict over the future of the nuclear industry in Germany. As Pascale Hugues, a French commentator put it, 'La menace n'est pas crédible dans cette idylle pastorale' (1998, p.91).

The conflict focused on Gorleben has lasted for nearly four decades. At its height during the opening years of this century its visible signs were never hard to find throughout the Landkreis (county) of Lüchow-Dannenberg, the administrative area covering the Wendland in which Gorleben is situated. The capital letter 'X' has become a potent symbol of the anti-nuclear resistance. 'Tag X' (Day X) was used as a rallying call during the long build up to the first transport of high level radioactive wastes into the region. Throughout the area on gateposts, on houses, in fields and by the roadside, yellow wooden 'Xs' have been erected, the ubiquitous symbol of protest in this region. Slogans such as 'Ausstieg' ('Climb down') and 'Stop CASTOR',[1] referring to the massive canisters which transported the waste

FIGURE 5.1 Aerial view of Gorleben and surrounding region

Source: GNS Gesellschaft für Nuklear-Service

FIGURE 5.2 Location map of Gorleben

Map by: John Hunt

into Gorleben, were daubed on walls and on the tall electricity sub-stations that are dotted around the countryside. 'Wir Stellen Uns Quer' (roughly 'We make our stand') announced the implacable determination of the protest movement. A bright orange sun symbol on a green background displayed on posters and flags proclaiming the 'Republik Freies Wendland' (the Free Republic of Wendland) gave a symbolic and potent identity to an area whose integrity was threatened. When I last visited the Wendland in 2014, three years after the last convoy of CASTOR flasks had violated this placid landscape running the gauntlet of massed protest, the yellow crosses were still there, evidence that, despite its victory in stopping further transports, the Gorleben protest movement was still active.

On a country road through the forest south of Gorleben is the source of this long and bitter conflict. On one side of the road is a guarded complex of office buildings and mine head workings. This is the site of the Gorleben salt dome in which a mine has been excavated for scientific exploration for a potential deep disposal repository for Germany's radioactive waste. This is still the only site so far in Germany identified for a potential deep disposal repository for high level wastes and spent fuel. The Gorleben salt dome, one of over 250 in northern Germany and containing 600,000m³ of salt, is 14km long × 4km in width and extends from 250m below the surface down to 3,400m. The excavations reach the exploration level at a depth of 840m and the repository, if constructed, would be excavated below this level. There are 7km of roads within the mine and about 1,500 measuring and monitoring instruments. Around a fifth of the underground exploration has been completed and so far, according to a Federal Ministry of Economics and Technology, BMWi report (2008), it has not revealed problems which would rule out the possibility of a final repository at Gorleben. This view is challenged by arguments that there is contact with groundwater at the site and that it is therefore unsuitable on safety

FIGURE 5.3 Symbols of protest

Source: Left: Fice 2010, Creative Commons; centre and right: author

grounds (Damveld and Bannink, 2012). Clearly there are key technical issues still to be resolved about Gorleben's suitability, including contact with water, existence of hydrocarbons, potential fracturing and problems of retrievability (Röhlig, 2013). Nevertheless, the Gorleben mine exists and, while it does, it is difficult for it to be ruled out as a site for Germany's geological repository.

Across the road from the mine and set within the forest is another set of buildings containing an interim store for vitrified high level wastes, a low and intermediate level waste store and a mothballed pilot conditioning plant where wastes were to be conditioned into a form suitable for final disposal. Gorleben was selected as the location for this interim store, in part for storing lower activity wastes from within Germany but also, and more controversially, to receive high level wastes sent back from reprocessing in France and Britain as well as spent fuel from Germany's power plants. No German spent fuel has ever been delivered to Gorleben and, since 2011, no further shipments of repatriated wastes have been allowed in. The store for intermediate and low level wastes is operational but the high level waste store is only a quarter full and the pilot conditioning plant designed to treat and package spent fuel for final disposal is effectively redundant.

The physical legacy of nuclear power at Gorleben, then, consists of an excavated but unfinished mine, a partly empty high level waste store and an unused conditioning plant – testament to the waxing and waning of the opposing interests that have fought over the Gorleben project for nearly two generations. And, on the edge of the forest beside the road leading into the mine, a ship, the *Beluga*, formerly used in protests by Greenpeace, was erected overnight and stands, now high and dry, as a silent and symbolic epitaph to the long struggle fought over this terrain.

Over the years both the salt dome and the interim store have each been the target of protests which have reverberated across the country. In the early years, during the 1980s and into the 1990s, it was the salt dome, as the potential final repository for Germany's high activity wastes, that attracted attention. At the turn of the century, in 2000 in response to protest, an anti-nuclear coalition at both federal and regional levels of government imposed a moratorium on further exploration at the mine which lasted until 2010. By that time the focus of protest was the interim store bringing to a halt the long established annual transport of CASTOR flasks of high level wastes from the reprocessing facility at La Hague in France. Exploration at the mine was briefly resumed in 2010 only for it to be suspended in 2012, in the wake of the Fukushima accident in Japan and the consequent anti-nuclear actions in Germany.

During more than three decades of action and activity the anti-nuclear campaign around Gorleben had become transformed into a social movement, with a continuing momentum and life of its own. The annual autumn protests were a social as well as political event. 'The CASTOR season was our carnival' said one of the local politicians I interviewed in 2014. To those who had been part of it from the beginning it seemed a bit of a miracle that the resistance had never been broken and that membership had continued to grow. It was difficult to realise that victory had been won, not only at Gorleben but, along with the wider anti-nuclear

FIGURE 5.4   Greenpeace vessel *Beluga* at Gorleben

Source: Author

FIGURE 5.5   Headworks of the salt mine at Gorleben

Source: Author

movement, against nuclear energy itself. As one of the original and most implacable members of the Gorleben movement, Wolfgang Ehmke, with four generations of his family participating, declared, 'The campaign against CASTOR may be over but we must stay on the alert. The fight against nuclear goes on. Our resistance has never been broken' (Wolfgang Ehmke, interview, 2014).

In terms of the three broad themes of this book, Gorleben is distinctive and different from the other three communities in several key respects. As to its nuclear legacy, the first theme, Gorleben lies at the opposite end of the continuum to Hanford having arrived on the scene much later. Moreover, in the sense of a nuclear community based around a significant nuclear complex as, for example at Sellafield, it has never arrived at all. The industry has never become established on anything like the scale that was originally envisaged, its imprint being confined to an underground laboratory, a waste store and a pilot conditioning plant. But, as a nuclear community in a social and political sense, Gorleben, like Sellafield, is a name associated with struggle over nuclear energy and its dangerous legacy. Throughout its history Gorleben has been unsettled and unstable, in contrast to Hanford or La Hague where the nuclear industry and local community exist side by side in a symbiotic relationship.

Considering the second theme, the power of discourses, Gorleben expresses a continuing conflict between opposing nuclear discourses rather than the discursive shift experienced at Hanford and latterly at Sellafield as they moved from production to clean-up. From the outset in Germany the discourse of Danger and Distrust has had a dominating presence shaping the power relations between a vigorous anti-nuclear movement and a seemingly resilient and implacable nuclear presence. Gorleben represents the politics of resistance, the persistent determination by highly motivated activists to prevent the transformation of a region into the permanent home for Germany's nuclear wastes. With the construction of the underground mine and the surface interim store the nuclear industry gained a foothold but not one firm enough to guarantee a continuing presence. Gorleben was the primary battleground between the Federal Government with its local presence and its antagonists where, ultimately, local tenacity and determination combined with broader anti-nuclear forces in German society as a whole, hampered and eventually halted the progress of the industry resulting in stalemate. Such an outcome was by no means certain and, in my earlier visits to Gorleben, was impossible to call.

In the case of the third theme, the power of the periphery, Gorleben demonstrates a community's determination initially to push back and ultimately prevent the further intrusion of the nuclear industry, for the time being at least. But, it is not simply a case of an apparent David beating back a mighty Goliath. The Gorleben movement was part, albeit a critical part, of a much broader coalition of forces that were eventually able to exert sufficient social and political power to overcome a weakening nuclear industry. Gorleben, at least in the early stages, exhibited the hallmarks of the periphery with all the vulnerabilities that implies. As time went on, the Gorleben movement was able to exert influence and leverage through its wider associations and linkages; the periphery became more and more central, part of a broader social and political landscape. The outcome was, as always in such conflicts, a product of changing power relations brought about by changing discourses and the ability to mobilise resources (Hinchliffe and Belshaw, 2003).

This chapter will consider the course of the conflict and its outcomes to try to explain why for so long the protagonists remained so unyielding and intransigent. The perspective is informed by several visits to Germany over three decades to discuss issues with those involved in government, in the nuclear industry and in civil society focusing especially on the community around Gorleben. The people I met and interviewed are listed at the end of the chapter. My argument proceeds as follows. Firstly, the history of protest in Germany is explored to identify Gorleben's prominence as the focus of conflict. Secondly, the nature of the conflict and a growing polarisation is identified. Thirdly, the way in which the conflict was shaped, sustained and finally resolved through the power relations of the main protagonists is discussed. This leads on, fourthly, to an explanation of the nature of the conflict in terms of the peripheral nature of the Gorleben area. Fifthly, the earlier attempts to find ways forward for achieving solutions to the problem of

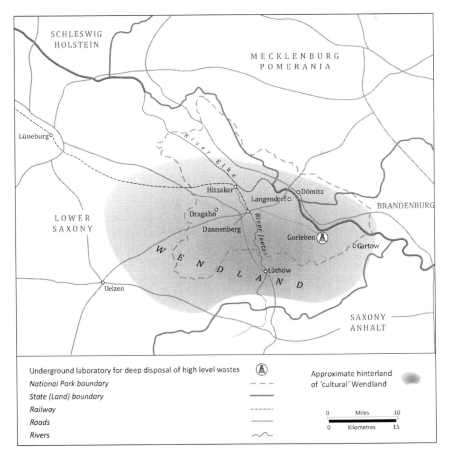

FIGURE 5.6 Map of the Wendland region

Map by: John Hunt

radioactive waste management within Germany are considered. Sixthly, the narrative explores the eventual triumph of the anti-nuclear movement in the wake of Fukushima and the opportunities it presents for a new approach to radioactive waste policy. This leads to a final section of reflection on the role of Gorleben in the continuing uncertainty about radioactive waste management in Germany.

A history of protest

The rise and fall of the Entsorgungskonzept

The selection of Gorleben as a site for nuclear waste management has its origins in the announcement in the early 1970s by the Federal Government of West Germany of plans for an *Integriertes Entsorgungskonzept* (Integrated Waste Management Concept). This would consist of a single location for a reprocessing plant, waste processing and conditioning facilities and a deep geological repository. In 1973 the Federal Government had initiated a siting strategy for a repository. Salt formations in the form of salt domes, abundant in northern Germany, were considered to offer the appropriate host rock for the repository. According to Detlef Appel (2006, p.57), a geologist,

> The main advantages were seen in the 'impermeability' and the dryness as well as the plastic, mechanical behaviour (creep, convergence) of rock salt, excluding the existence and flow of groundwater and preventing the formation and long-term existence of open fractures.

The Federal Government identified three locations in the Land of Lower Saxony for comparative evaluation (near Oldenburg, Celle and Hannover respectively). Local opposition was aroused and in 1976 the investigations at the sites were halted. Meanwhile, the Land government of Lower Saxony which was in favour of the project decided to undertake its own selection process. In 1976 Ernst Albrecht, the Christian Democrat (CDU) premier of Lower Saxony, indicated the selection of Gorleben as a site for exploration and it was formally announced early in the following year and later accepted as a suitable site by the Federal Government. Thus ended any prospect of a comparative siting process and Gorleben appeared as the exemplar of the Decide Announce Defend approach that was characteristic of siting policy at that time. In the words of Asta von Oppen, an anti-nuclear protester in the Wendland, 'Albrecht opened the door and in came big industry' (interview, September 2003).

The reasons behind the selection have never been fully revealed. According to Peter Ward of DBE, the company that manages the salt dome, 'No one knows the real reason why Gorleben was chosen in the first place' (interview, 2014). Bruno Thomauske, of the Federal Office for Radiation Protection, confessed, 'For me it is astonishing that the documents couldn't be found' (interview, May 2003). Anselm Tiggermann (2010) believes the lack of comparative data suggests political motives played a role. On the other hand, a detailed report by the Federal Ministry of Economics and Technology (BMWi) stated that 'the site was selected in a highly scientific and methodical process' (BMWi, 2008, p.16). The report argues that the Land government used extensive criteria, scientific, socio-economic and environmental, to reduce a long list of sites down eventually to Gorleben which best met the criteria. It goes on, 'There is no foundation to any accusations claiming the selection process was influenced by policy makers which could have resulted in Gorleben being "preselected" as the site' (p.19). Moreover, it makes the further claim that, 'In retrospect, it can be said that the Gorleben project was revolutionary in terms of the steps it took when communicating with the public' (p.18). There does appear to have been a combination of scientific reasons including the presumed suitability of the salt dome and social criteria notably the low population density and, perhaps, a perception that the jobs and wealth the investment might bring to the area would be welcomed, but they were not specific to Gorleben. Its location right on a heavily guarded border with the former East Germany (German Democratic Republic) might seem provocative and would appear, on the face of it, to be disadvantageous especially for site investigations. But, the reservations of the Federal Government were not believed to be serious and, in choosing Gorleben, Prime Minister Albrecht had asserted the role of the Land in the site selection process.

The selection of Gorleben without public engagement or explicit comparative assessment was, to say the least, disingenuous. At the time, consultation was with the local councils as representative government; local action groups were in their formative stages and tended to adopt a hostile stance from the outset. It was unclear why Gorleben had been chosen in preference to other sites that were at least as technically suitable. From the Federal Government's perspective having a single site would confine the opposition (though they could not have anticipated just how resolute it would be) and divert it to the Land government. But, the lack of transparency about the selection of Gorleben, then and later, fuelled a sense of outrage and was 'a major reason for the distrust of many people in political decision making in the field of radioactive waste management in Germany' (Appel, 2006, p.59).

The scale of the protests that followed the identification of Gorleben, though massive, was not unprecedented. This was a period in which there were growing and occasionally violent protests. Indeed, the anti-nuclear movement in Germany had begun in the 1950s first as mass protests against the deployment of nuclear weapons on German soil. It was well established by the time the first nuclear power stations were under construction. Opposition to nuclear power had become the new environmental movement's focal point (Dryzek et al., 2003). Despite the vigorous promotion of nuclear energy as an efficient and safe answer to the country's future energy self-sufficiency, proposals for power plants were strongly opposed by some local communities which 'reacted as if they had been handed a rattlesnake' (Dominick, 1992, p.167). In some places they were successful, if only for the projects to be imposed on other communities less able to resist. But, in every case, the experience 'showed how the intrusion of the nuclear menace could mobilise middle-of-the-road and even conservative elements for anti-nuclear action' (ibid., p.163).

The strength of the movement against nuclear power lay in its origins in the radical protest movements of the 1960s and 1970s with wider linkages to other social issues – the peace movement, environmentalism, gender inequality – that gave breadth and diversity to its membership. This was the period which inspired the formation of the German Green Party (Die Grünen, in 1980), Greenpeace Germany and other radical groups. It also saw the formation of the citizens' initiative groups under the umbrella of the Federal Association of Citizens' Initiatives on Environmental Protection (Bundesverband Bürgerinitiativen Umweltschutz, BBU) that came to play such a prominent role in the Gorleben protests. Not only was the anti-nuclear movement large, certainly much larger than in any other west European country, it was also, and continued to be, more confrontational than elsewhere (with the exception of France see Chapter 4). A singular reason for this was the lack of opportunity for civil society to participate in policy and decision making. Detlef Appel (2006) notes the tendency for closed decision making over a long period during which 'there was no attempt to a broader information or even participation of the public' (p.61). This was also evident in other countries at that time but Germany did not share fully in the opening up of decision making processes to non-institutional and non-expert so-called 'stakeholders' that, as we have observed, became characteristic of the 'participative turn' during the latter years of the twentieth century. John Dryzek and his colleagues, in their masterly analysis of social movements in four countries, describe the German case as that of 'passive exclusion', a system of governance that relies on institutional expertise (bureaucratic, academic and independent research organisations) for policy formulation leaving it up to wider participation from civil society to address the revealed policy. As they put it, 'The environmental movement in Germany therefore encounters passive exclusion in which opportunities for formal political inclusion are limited and unconventional challenges to governmental authority have been strongly resisted' (Dryzek et al., 2003, p.41). That is, up to a point, as we shall see. If influence over policy making is denied then it leaves the way open for a policy that is widely opposed to be ultimately overthrown through mobilisation of opposing power precipitated by a changing discourse.

By the early 1970s anti-nuclear protests were overshadowing all others in terms of the numbers and breadth of the coalitions that were mobilised. During these years protest was against nuclear power plants that were coming on line. At its peak Germany had 39 nuclear reactors which by 2011 had become 17 at 12 sites producing around a quarter of the country's electricity, declining to around 18% by 2014 with a final shut-down of all nuclear plants to be completed by the end of 2022 (Blowers et al., 2008; Damveld and Bannink, 2012; IPFM, 2011; WNA, 2014). One of the biggest protests was against a proposal for a nuclear power plant at Wyhl in southwestern Germany. It sparked what Koopmans has called 'the strongest and most persistent anti-nuclear movement in the world' (1995, p.158).

FIGURE 5.7 Map of nuclear sites in Germany

Map by: John Hunt

Here, local farmers were joined by student activists and 20,000 people were involved in the occupation of the site for eight months which contributed to the eventual abandonment of the project in 1977. Protests, starting in 1976 over the Brokdorf power station on the Elbe downstream from Hamburg, attracted large numbers (figures of 30,000 up to 45,000 have been estimated) and clashes with police signalled a more confrontational form of conflict that characterised subsequent protests. Another characteristic was the use of legal procedures to change or delay decisions and, in this case, the courts ruled in favour of a moratorium on construction delaying its opening until 1986. At Kalkar on the Dutch border, in 1977, there were mass protests by 60,000, including Dutch people, against the fast breeder reactor which was abandoned for technical and commercial reasons some 20 years later. There were protests, too, at Grohnde in Lower Saxony. The 1970s was a high point of anti-nuclear protest in terms of the sheer numbers involved at several specific sites. But, the ability of the anti-nuclear movement to mobilise large numbers did not diminish. And the biggest of all were the protests against the proposals for a reprocessing and waste management complex at Gorleben.

Much was at stake for both sides in the conflict at Gorleben. For the nuclear industry the project would deal with the back end problems of energy production, the need for a reprocessing plant in conformity with the federal Atomic Law which required a closed nuclear fuel cycle, and for a final repository which would deal with all the country's accumulating high level nuclear waste. Elsewhere in Lower Saxony, at Salzgitter, the Schacht Konrad project for the deep disposal of non-heat generating (intermediate and low level) wastes in an abandoned iron ore mine at a depth of over 1,000 metres was proceeding slowly despite legal impediments. Thus the two significant projects, Gorleben and Konrad, together would provide a concentration of disposal facilities for all types of radioactive waste within a relatively small area in the centre of the country. However, two other projects in the area were to prove unsuccessful and faced premature closure. At Asse, also in Lower Saxony, drums of low and intermediate level wastes were stored in deteriorating conditions and would eventually have to be retrieved while, across the border in Saxony Anhalt in the former East Germany, the Morsleben low level waste repository also in salt was being decommissioned.

From the perspective of the anti-nuclear movement, the Gorleben project would present a solution to the problem of nuclear waste and thereby give the green light to the further expansion of nuclear energy. It was, therefore, vital to prevent it since, in Germany, the fate of a nuclear power station's operating permit was tied to progress in the repository projects at Gorleben and at Konrad. According to Wolfgang Rüdig, 'Gorleben thus came to symbolise the megalomaniac plans of the nuclear industry, envisaging the construction of dozens of further reactors and the preparation of the transition to a plutonium-based "all nuclear" energy future' (2000, p.50). Among the Gorleben protesters in the early years were some who had been involved in protests elsewhere, notably at Wyhl and, more significantly, at Langendorf a few kilometres from Gorleben where a nuclear power plant on the Elbe had been proposed.

Protests at Gorleben began as soon as the projects were announced, gathering crowds of up to 20,000 people. Local leaders emerged who would provide the movement's dynamism and identity in the years to come. Among them was Lieselotte (Lilo) Wollny who, for four years during the 1980s, was a Green MP in the Bundestag. Another was Marianne Fritzen, described as a 'moral institution', who held the movement together. Following on has come a younger radical generation including Rebecca Harms, a leading Green member in the Lower Saxony Parliament and later, as MEP, President of the Green Group in the European Parliament but maintaining her close links with the movement. There was also Wolfgang Ehmke, its chief spokesman in later years. But leadership was also given from the outset by Graf (Count) Andreas von Bernstorff (and latterly by his son Fried) based at his Schloss in the large village of Gartow and by Pastor Eckhard Kruse of the Lutheran Church in Gartow, adding the landed aristocracy and the church to the broad coalition that comprised the Gorleben movement. Together these leaders with several others formed the 'Gartower Runde' (Ring or Round Table), a rather elitist core group that provided the cohesion and coordination of the Gorleben movement.

During 1978–9 technical hearings were held in Hannover to appraise the integrated reprocessing, storage and disposal project. The Gorleben International Review attracted over 100,000 people (some estimates put the figure as high as 140,000), the biggest ever anti-nuclear demonstration up to that point, many of whom had walked or driven tractors all the way from the Wendland to be present. The hearings took place against the backdrop of the accident at Three Mile Island fuelling the anti-nuclear sentiment that was abroad. Following the hearings and

FIGURE 5.8 Anti-nuclear protests at Bonn in October 1979

Source: Hans Weingartz

protests, Lower Saxony Prime Minister Albrecht withdrew the proposal for reprocessing on 'political' grounds as not suitable for Lower Saxony though he accepted it was technically satisfactory. The concept of an integrated nuclear waste management centre was thus abandoned. Far from mollifying the protesters, the withdrawal of reprocessing from the Gorleben site merely concentrated their targets on to the interim store and the attendant repository.

The triumph of protest

For a while the search for an acceptable site for reprocessing continued. Another site in the area at Dragahn near Dannenberg was among several briefly considered a few years later in pursuit of the integrated concept split between different locations. The eventual choice for the reprocessing part of the concept fell on Wackersdorf in rural Bavaria not far from the Czech border. Here, too, there was trenchant resistance and mass protests by farmers and local activists supported by environmentalists. The economics of reprocessing had also begun to look unfavourable. This proposal was also eventually dropped leaving Germany to rely on France (La Hague) and Sellafield (UK) to provide the reprocessing facilities that enabled the utilities to fulfil the requirements of the Atomic Law for a closed nuclear fuel cycle. It should be noted here that this requirement was dropped when the Law was amended in 1994 to allow nuclear utilities the option of storing spent fuel for ultimate direct disposal as an alternative to reprocessing. This was a pragmatic as well as political response to the problem of finding a site for reprocessing in Germany at a time when reprocessing itself was falling out of favour for commercial and technical reasons.

Even the option of reprocessing was removed in the wake of the election of the Red/Green Federal Government in 1998, the first time the Greens had participated in government. Both in coalition with the SPD at Länder level and at federal level Die Grünen have continued to push hard on the nuclear front (Rüdig, 2000). As a result of Green participation in the Federal Government a set of proposals emerged which were incorporated in a new Atomic Energy Act of 2002. These included: a phase-out of nuclear power; the abandonment of reprocessing; the construction of interim spent fuel stores at power plants; and a review of nuclear waste policy which included the suspension of further exploratory work at the Gorleben mine for between three and ten years. Although contested, over the next decade these proposals eventually became accepted policies.

There were four agreements arrived at through the consensus and embodied in the Atomic Energy Act of 2002:

1   The agreement to a phase-out enabled nuclear plants to continue to operate for a maximum of 32 years of electricity output with flexibility to switch production from one plant to another. In effect, this meant that some plants might operate well beyond 32 years and the flexibility to switch would allow companies to concentrate on the most efficient plants rather than invest in

maintaining the less efficient. An amendment to the Atomic Energy Act in 2010 extended the lives of operating plants. This prevarication was swept away in the post-Fukushima settlement of 2011 which ushered in the *Energiewende* (Energy Transition) including the accelerated shut-down of all 17 of Germany's nuclear power plants, eight immediately and the rest between 2015 and 2022 (Völzke, 2014; IPFM, 2011; SRU, 2011).

2   The second agreement of the 2002 Energy Act related to the phasing out of spent fuel transports to reprocessing facilities by 2005 although it was made clear that existing contracts with COGEMA (now Areva) in France and BNFL (now the Nuclear Decommissioning Authority, NDA) in Britain would have to be fulfilled. This, too, was acceptable to the industry which had already moved its operations from reprocessing to a once through cycle with direct disposal of spent fuel.

3   As a consequence the third proposal under the Energy Act was the development of spent fuel stores at each power plant in order to avoid transports within the country. Although this provoked anti-nuclear protests at various power plants it avoided the bigger problem of confrontation and disruption that would inevitably accompany internal transport of spent fuel to Gorleben or Ahaus (an interim store for spent nuclear fuel in the northwest of the country). Future transports would be confined to shipments from France and Britain to Gorleben. But, as a result of the protests, shipments of repatriated wastes to Gorleben also ceased, leaving 113 casks in a store with a 420 cask capacity and 26 casks (21 from Sellafield and 5 from La Hague) due to be returned but with nowhere to go (BfS, 2014; Völzke, 2014; Rüdiger Kloth, interview, 2014). However, in 2015 four interim storage sites at nuclear power stations were identified to take the casks: Philippsburg for the five from La Hague; and Biblis, Brokdorf and Isar for the shipments from Sellafield.

4   The fourth element of the consensus concerned the long-term management of radioactive waste. Further investigation of the Gorleben salt mine was suspended for at least three but no more than ten years beginning from 1 October 2002 while a full review of options was undertaken. In fact it went the full term. The mine would be fully maintained and monitored during this period. From the nuclear industry's point of view their investment was protected although the power of the anti-nuclear movement had halted further progress on the project. Meanwhile, with high level wastes being managed overseas in Britain and France or at interim stores at the power plants, there was no great urgency to find a long-term solution. In addition there was the prospect of a deep repository constructed in the former Schacht Konrad iron ore mine also in Lower Saxony near Braunschweig shortly becoming available for non-heat generating wastes.

In practice, it may be said, the first three proposals, forged in consensus talks between the nuclear industry and government, reflected a compromise in which the industry yielded very little beyond what was commercially acceptable (Rüdig, 2000). This was rather less true of the fourth proposal. The industry achieved a temporary reprieve under the CDU/FDP (Conservative/Liberal) Federal Government which was formed in 2009 supporting life extensions to the power plants and lifting the moratorium on exploration of the salt mine (IPFM, 2011; Damveld and Bannick, 2012). Work resumed in 2010 to 2012 but only briefly and, for part of that time, the site was virtually under siege from protesters blocking access, damaging cars and getting into the site itself. Peter Ward, geotechnician and also chairman of the workers committee at DBE, the site's operator, spoke to me of the tensions and fears of the workers as attempts were made to prevent them getting to and from work with the police standing off while workers were bussed in to avoid being stopped (interview, 2014). Once again, exploratory work at the mine was stopped, this time including monitoring. Various options for its final closure have been explored including flooding it and enclosing it in concrete. A Nuclear Waste Management Commission to define a fresh siting process starting with a 'white map' and all options open was set up in 2013. For the present the Gorleben mine must be kept open until it is eliminated, if it ever is, in a comparative siting process. Although the Gorleben mine has so far survived, its status is ambiguous and its future uncertain.

The vicissitudes of Gorleben's history reflect the changes in power relations over nearly four decades that have shaped the political context in the struggle both over Gorleben and in nuclear policy writ large in Germany. While the discourse of Danger and Distrust has been dominant throughout there have been shifts in the power relations and in the resources that the contending interests have been able to deploy at different times. The result has been the physical presence of a potential underground repository that is in limbo and an interim waste store closed to further shipments of high level wastes. For long the outcome seemed uncertain as the battle raged to and fro. While a foothold was established by the industry it has never proved to be the harbinger of an enduring presence. But, so long as the potential of these projects remains unfulfilled there is the incentive for the combat to continue. In any case, despite their triumph, the protesters are loath to let go. For them, if Gorleben exists then it must be an option, possibly the only option. Equally, for the nuclear interests, if Gorleben does not exist then their best hope for a site for a deep repository disappears.

The nature of the conflict

Opposing the mine

There has been sustained opposition to both the mine and the interim store, keeping Gorleben at the front line of the anti-nuclear protest movement (Blowers and Lowry, 1997). In the early years, the salt mine was the focus of attention indicating the range of methods used by the protesters. In 1980, shortly after the Hannover demonstrations, protesters erected a temporary village called 'The Free Republic of Wendland' complete with its own flag, pirate radio station and 'passport' (which the Graf von Bernstorff, for one, still keeps), an action condemned rather humourlessly as high treason by the Lower Saxony Minister of the Interior. Also in 1980 the site of a prospective deep borehole was occupied for 33 days. Since then there have been numerous protests, mostly peaceful but some more violent. There have been actions that have intimidated and threatened those working at the site, including destruction of property and dropping things down the shaft when miners were underground. The peaceful protests have adopted varying tactics including rallies and church services in the woodland, concerts, political theatre, marches, sit-ins and local group activities focusing on raising awareness and fund raising. Weekly prayers are still held each Sunday in the woodland just outside the site. These activities have been accompanied by the characteristic outpouring of pamphlets, petitions, stickers, T-shirts and slogans, the daubing of graffiti, the use of the media to promote the cause and the lobbying of decision makers. The style of protest has been inventive, carefully prepared but also spontaneous. For example, one protest I attended in 1995 centred on planting trees on the railway lines used to transport the flasks. At another, which I also witnessed, in 2003, protesters hung large photographs of babies and small children round their necks proclaiming the stark slogan, 'Unsere Kinder soll lachen nicht strahlen' ('Our children are supposed to laugh, not radiate').

One of the more imaginative actions was the creation of the Salinas Company to exploit the resource of the Gorleben salt dome. The company was set up to mine salt under land owned by Graf von Bernstorff, as an alternative to the proposed repository which overlaps the count's land. The real purpose was to test the law since the Federal Office for Radiation Protection (BfS) which is responsible for nuclear waste management had appealed against the licence granted by the Land to mine the salt. The project was a half serious attempt to determine which is the most appropriate use for the salt dome – research for radioactive waste disposal or the production of salt. 'We want to use the salt – if it happens to stop the project then we can't help that' commented Thomas Hauswaldt, managing director of Salinas (interview, 2003). The salt (actually from Göttingen) is sold in packets bearing the Wendland orange and green motif. An alternative view of the

FIGURE 5.9 Peaceful protests at railhead, Dannenberg, 1995 (left), and at Gorleben, 2003 (centre and right)

Source: Author

FIGURE 5.10   Salt from the Salinas Salt Company

Source: Author

salt project is that it was a deliberate attempt to destroy the potential of the site under the pretence that it was using the resource for a product knowing there was no serious market for it.

Confronting CASTOR

On the whole the early protests against the mine, though large in scale and with large police deployments, were broadly peaceful and coordinated through the local Gorleben movement. By the 1990s when attention shifted to the interim store and the CASTOR transports, the style of protest became more confrontational with the occasional protests organised from outside the region but seemingly tolerated by the moderates. These protests employed more destructive and violent tactics such as undermining and blocking roads, sometimes with steel wires at head height, infiltrating railway property and bridges and carrying on running battles with police. The first shipment of spent fuel came in April 1995 from the Philippsburg power plant in southern Germany. It was met by blocking railway lines and there were clashes between police and protesters. The first shipment from La Hague came in May the following year and, once in Germany, it was delayed in the final stages of its journey by bonfires, sit-ins and blockading by farmers using tractors and dumping manure (potatoes have also been a favourite barricade). The police met the protests with water cannon, tear gas and batons. An even bigger protest against the transports came in March 1997 and accompanied the shipment of six casks from La Hague throughout its three day rail journey in Germany. It culminated in mass demonstrations especially along the final stage of the route by road from Dannenberg to Gorleben. Although most of the protest was peaceful and passive the violence was alarming. Protesters pelted police, disrupted the railways, cemented themselves to the track, undermined the road and blocked the convoy with tractors. It took a massive deployment of around 30,000 police and border troops armed with water cannon, riot gear, helicopters and a tank in one of the country's largest internal peacetime operations eventually to secure the passage of the casks into the store (Blowers and Lowry, 1997). The estimated cost of the operation was £25 million.

In the following years the protests continued against the annual transports from France which were switched from the Spring to November in a vain attempt to deter opponents. In 2001 there was another big protest with damage done to a railway bridge and skirmishes with police. The protests continued every year until 2011, the final year of transports into Gorleben. The authorities appeared to regard the transports as having a diminishing impact. According to an official report, the 2002 transport of 12 CASTOR casks 'received a high media coverage, but disturbances by demonstrators were minor' (BMU/BfS, 2002). Over the following years until the mass anti-nuclear protest of 2010 (see below), the numbers of protesters involved diminished to around 5,000 but the police presence was still formidably large, as high as 12,000, a mixed force of local and border police and those brought in from further afield, notably Berlin, Hamburg and neighbouring Länder such as Sachsen-Anhalt and Mecklenburg-Vorpommern. Such a force deployed to deter violent actions had to be armed and equipped, requiring vehicles and containers for sleeping. In this sparsely populated region (Lüchow-Dannenberg Landkreis has about 50,000 inhabitants) the police presence in their green uniforms is highly visible. The Wendland is 'the only region in Europe that becomes green in Autumn' (interview with Jürgen Auer, Public Relations Manager of BLG, now GNS, the company running the interim store, 2003). 'In November, everywhere the leaves have fallen. But, in our forests the leaves are still green – there are so many police' (Thomas Hauswaldt, interview, 2003). Such force may seem excessive to subdue protests that are overwhelmingly peaceful and include large numbers of

FIGURE 5.11 Confrontations between protesters and police

Source: Top left and right: Michael Mariotte; bottom left and right: author

elderly people and young children. There was a feeling of intimidation and growing anxiety which spread during the days before the transports reached the Wendland. Images of armed state power crushing unarmed protesters by overwhelming force have provided a potent source of sympathy and support for the anti-nuclear cause. 'People get frightened when the CASTOR comes with the helicopters, dogs and horses' (interview with protester Asta von Oppen, 2003).

The heavy-handed approach is justified on the grounds that, in the face of potential violence and disruption and the possibility of terrorism, the security of such dangerous cargoes is paramount. Thus, fear characterises both sides in the conflict. There is the fear on the part of the authorities, expressed by Jürgen Auer that, 'Even a small number of demonstrators could do harm' pointing out that it only took four people to undermine the railway line to Dannenberg. The police are criticised for not having a consistent strategy and failing to intervene when sites are invaded and workers impeded. 'The protests have a lot of approval in a large proportion of society. The end justifies the means. It's cool to be against Gorleben' (Peter Ward, interview, 2014).

There is little doubt that over the years the conflict over Gorleben became more confrontational. But, the emphasis on peaceful means of protest persisted among the local leaders. For instance, Marianne Fritzen recalled that the ideas and methods inculcated in the cities in the wake of the radicalism of the late 1960s was not welcomed in the Wendland. It is difficult to define when the boundary of violent protest is crossed. 'What is violence?' asks Wolfgang Ehmke. 'Is sitting on a road or railway violence?' (interview, 2003). The threshold may be crossed when safety is compromised or lives are threatened as, for example, when trains are derailed or roads undermined. There are elements among the protesters who believe in and practise methods of civil disobedience. A small NGO, *Kurve Wustrow,* Centre for Education and Networking in Non-violent Conflict Transformation, based in the area, provided training in non-violent forms of civil disobedience. Forms of direct action, occasionally including violence, undoubtedly contributed to the escalation of the conflict and the more robust response from the authorities. According to the protesters, a consequence was the incremental suppression of civil liberties. They claimed the Land government of Lower Saxony provided the police with extensive powers to curb the protests. Protesters were detained without access to lawyers; sometimes they were transported from the area and dumped in woodlands. There were accusations of phone tapping, small groups were often banned and, during the transports, protesters were kept away from the road. There were also claims of police infiltrating the protesters. Whatever the extent or truth of all this, it suggests a condition of polarised and uncompromising conflict.

Wolfgang Ehmke reflected on the history of protest in the region in an interview I had with him in 2014.

> Our struggle is non-violent. We did not attack people and policemen. We are proud of our non-violent movement. There were sometimes situations where the police were violent and in later years there was some violence. Much of the action was symbolic. Our resistance has never been broken. It is a little bit of a miracle that we have struggled on for more than a generation.

Over that time there has been a slow but inexorable and ultimately decisive change in the nuclear discourse and a concomitant shift in power relations as the anti-nuclear interests have achieved success.

The power of opposing interests

Anti-nuclear protesters

What accounts for this transformation of power relations and the success not only of the Gorleben movement but also of anti-nuclear forces in Germany as a whole? The explanation lies, I think, in the ability of protest to reflect and to shape the discourse, capturing the essence of anti-nuclear feeling and anxiety and proselytising a set of views and values that eventually become mainstream. Thus, in Germany at least, anti-nuclear has become over time the normative position and pro-nuclear increasingly marginalised, defensive and defeated. As this shift has occurred so the resources available to the anti-nuclear discourse have increased to the point where it has become possible to influence and then dominate the policy making process. And the Gorleben movement has played a crucial role in all of this.

Gorleben has been the fount and main focus of anti-nuclear protest in Germany for nearly four decades. According to Heinz Smital of Greenpeace Germany, the movement in Gorleben undertook the important work of integration, mobilising and developing the spirit of protest (interview, 2014). Although the scale and style of actions have changed, the significant feature of the protest has been its persistence, the ability to mobilise support for a campaign sustained over decades in changing political circumstances. What groups constituted this movement that shaped the conflict over Gorleben and whose influence spread widely across the country? What resources of power were they able to deploy?

Although, it must be said, the majority of the population were passive, non-participants in the issue, the Gorleben movement was able to mobilise resources of power through the breadth and diversity of its support and the variety of its contacts and methods. The movement drew together in loose liaison traditional and radical elements with a strong local base able to draw on wider support. The local activists were recruited from three distinct bases. First, there were the local citizens, many of them members of the Bürgerinitiativen (BI), part of the network (Bundesverband Bürgerinitiativen Umweltschutz, BBU) originally sponsored by the SPD (social democratic party) 'as part of its effort to expand citizen participation in politics' (Doherty, 2002, p.41). The BI had a formative influence on the environmental and anti-nuclear movement in the 1970s and eventually developed into a network of protest around the country. The Lüchow-Dannenberg BI has been devoted to the nuclear issue, first as an informal organisation opposing the proposed Langendorf power plant and, since its formal incorporation, as the focus of anti-Gorleben protest. Although it has elections and a constitution it manifests spontaneity. Through the variety of activities described earlier it provides the consciousness raising, networking, organisation and the ideological basis of the movement. It has about 1,000 members of whom around 10% form the activist core.

Secondly, the farmers' organisation (Bäuerliche Notgemeinschaft) forms a more traditional element of the movement but it is one that is integral to its success. At the outset the then nuclear reprocessing and waste management company (DWK) bought the forest at the mine site and some farmers rented the mining exploration rights to the nuclear industry. One landowner who refused to sell out despite the pressure and the large sums on offer was the Graf Andreas von Bernstorff, a crucial decision since he owns a third of the land covered by the mine's footprint. Farming opposition is strongest in the market gardening areas. The farmers' motivation is partly the fear that the image and, therefore, the market for their produce will be affected by association with radioactivity. But the idea of stewardship of the environment, that 'moral duty to look after and to hand it on in good order to future generations' (HMSO, 1990), was also strong, certainly with the Graf von Bernstorff. The farmers lent a more conservative respectability to the movement and thus broadened its base of support. They operated independently but, when necessary, could deploy the blockades of tractors, crops or manure which were so effective in hampering the nuclear transports in the Wendland. In Rebecca Harms's opinion, 'If the movement lost the farmers then it would be the end. People love the romantic connection' (interview, 2003).

The Gorleben movement also drew support from a third group, those from further afield, from the cities especially Hamburg and Berlin. Some, like Wolfgang Ehmke, were Wendlanders working in the cities but maintaining their links in the area. There were, too, those, especially Berliners, who bought second homes in the Wendland when it was the nearest part of the former West Germany. And there were those city people who have come to live in the Wendland. They are often professional people, teachers, artists, writers, some of them like Asta von Oppen, radicalised in the late 1960s, who brought different perspectives to the campaign. The strength of the movement lies in its ability to mobilise and motivate large numbers of people from across the social spectrum.

In addition to these local bases, the Gorleben movement was able to draw on external support coming from the national anti-nuclear groups, particularly Greenpeace, Friends of the Earth, BUND (Bund für Umwelt- und Naturschutz Deutschland, Association for Environment and Nature Conservation or Friends of the Earth, Germany) and Robin Wood. These NGOs have been effective in mounting demonstrations and blocking transports en route to Gorleben. For these groups, especially Greenpeace, Gorleben has been a useful target providing publicity and momentum in their pursuit of broader anti-nuclear agendas. They have tended to avoid a local presence which might antagonise more conservative elements in the protest movement and so prove counter-productive.

The Gorleben anti-nuclear movement derives its power from a variety of sources. It has achieved a proficiency in local organisation able to deliver variety and flexibility of response. It has proved able to mobilise support which cuts across political and social barriers. And it has been able to appeal to a wider movement which has the capacity to develop a powerful anti-nuclear discourse. In turn it has supplied a focus of energy, commitment and values that helped to inspire the broad, cross-cutting anti-nuclear protests that overwhelmed the nuclear industry in Germany in the post-Fukushima years.

Pro-nuclear interests

By contrast, the pro-nuclear forces have, over the long years of the conflict, been mainly on the defensive, retreating to a point where even the survival of what little remains of their Gorleben projects is in doubt. They are composed of three groups. First are those working in the industry or who benefit indirectly from its presence. The workers have felt threatened especially during periods of mass demonstration.

The occupation of the site by Greenpeace protesters while miners were still underground was felt to be positively dangerous for the miners. Many of the skilled workers needed were recruited from other parts of the country and complained of hostility and sometimes intimidation. They resented the bias of local media and the help given from time to time to protesters by the Land government though the protesters also complained of pro-nuclear bias at certain times. In addition there has been the insecurity caused by job losses. Consequently, workers express feelings of wariness like footballers 'coming out onto a playing field where the opposing team has been playing for some time' (interview with mineworkers, 1995). The insecurity at the mine was intensified by the moratorium, then by the brief resumption of exploration before the shut-down of 2012. With a declining workforce (down to 120 in 2014) and most of the workers on short-term contracts there was a pervasive air of pessimism among both management and workforce.

A second area of support for the industry, right from the outset, came from within the local councils. Following the abandonment of the reprocessing project, local mayors and councillors were persuaded by national leaders to accept the storage of nuclear waste, both in their own and the country's interests, at least as an interim measure and on condition there was no reprocessing for ten years (Federal chancellors Helmut Schmidt and Helmut Kohl visited the area to press the case). Local leaders were taken on foreign visits leading one protester to comment, 'These little mayors never saw Hamburg but now they have been somewhere, to Sellafield and La Hague. So, they think, this must be good' (Marianne Fritzen, interview, September 2003). The pattern of support is mixed, though. Support is strong in the *Gemeinde* of Gorleben in which the mine and interim store are situated and which receives annual payments from the Interim Store (€590k in 2014). The *Samtgemeinde* (combined communities) of Gartow, which also receives financial benefits, also tends to support the nuclear presence in the Wendland but with a small majority. As Bürgermeister Friedrich-Wilhelm Schröder commented to me: 'Opinion is divided – opponents are very loud, some don't care and supporters are not very vocal. I'm always for but always elected' (interview, 2014). Once outside the immediate vicinity of Gorleben the support fades and the Landkries (county) of Lüchow-Dannenberg is opposed. Support tends to follow political lines with the conservative CDU and the social democratic SPD usually in favour with the Greens and Independents (some elected on an anti-Gorleben platform) against though the issue crosses party lines. Supportive council members point to the economic benefits of the projects which are 'good for the area and needed by the nation' (interview with local mayors and councillors, 1995). By 2014 there was a sense of embattlement, a feeling of being abandoned by the politicians in Berlin and undermined by opponents. 'The region has been weighed down for 35 years and we'd like to have something to show for it. Opponents don't care where it goes as long as it doesn't come here' (David Beecken, county councillor, 2014). Like the nuclear workers, these local politicians feel opponents use fear as their weapon, intimidating those who support the mine. They point out that when the Integrated Concept was proposed it secured political support at all levels: local

(Gemeinde); county (Landkreis); regional (Land); and national (Federal). With changing political control during the 1980s and 1990s first the Landkreis and later the Land moved in opposition to the projects. Later the positions shifted again as the Federal Government came under the SPD/Green coalition in 1998 whereas the Land reverted to CDU control and a pro-nuclear stance after the elections of 2002. Following political shifts at national level the Gorleben moratorium was imposed, then briefly lifted and then again the mine was closed. Throughout all this political uncertainty there has been support for Gorleben from the councils but their influence is weak in comparison with the Federal and Land governments which have the political power to decide the fate of Gorleben.

The third arm of support for Gorleben comes, predictably, from the nuclear industry itself, both from its presence in the area and from its position in the country as a whole. But, in recent years both its presence and its position have diminished. In past times, the industry promoted its cause through the information centre in Gorleben through publicity material, exhibitions and educational programmes and through visits to the facilities themselves. Above all, the industry provided, and still does, financial support both directly through the so-called 'Gorleben Gelder' and through the local taxes it pays which help to underpin the local economy and infrastructure. This income is obviously attractive to local leaders in an economically depressed region. The tax and investment from the industry have provided the local area with improved facilities. As a result, Gorleben has created indoor and outdoor sports facilities while at Gartow development of a swimming pool and an artificial lake has enabled the village to market itself as 'Gartow am See', a small resort offering swimming and horse-riding, bird watching and cycling in the surrounding forests and heaths. Although the financial compensation in respect of the interim store continues to be paid, the industrial influence has been weakened, perhaps fatally, by the ending of CASTOR transports and the suspension of activities at the mine. The workforce is haemorrhaging, visits and publicity have ceased. The information already gained serves little purpose in the context of a new siting strategy although Gorleben is not ruled out. Bories Raapke, Managing Director of DBE which runs Gorleben, reflected on the situation in an interview in 2014:

> Gorleben exists and it should stay in. In the process of site selection it may be that the decision falls back to Gorleben one day. A final repository for radioactive waste is the most exciting environmental issue in Germany today. The nuclear industry has supported Gorleben in the past, put money in. Now industry is fading and there's no more money to be made. So, where is the money to keep Gorleben going? Industry won't want to pay especially as site selection and further sites will cost them. Politically Gorleben is unloved.

The pro-nuclear camp in the Wendland draws together a loose combination of industry, workers, scientists and politicians. The political and economic resources they are able to deploy have, so far, ensured the continuation of the Gorleben projects. In particular the resources already sunk in the mine and the interim store represent assets which the industry will not readily forgo. The economic costs of an alternative, let alone the political problems of finding a repository site elsewhere, make it unlikely the Federal Government will give up entirely on Gorleben.

During the long course of the conflict both sides have exerted their power with effect. In the end the anti-nuclear forces may have prevailed. For long focused on Gorleben the movement was largely responsible for the withdrawal of plans for reprocessing, the stopping of the CASTOR convoys and the suspension of the Gorleben repository programme. It was able to respond to and influence a shifting nuclear discourse and to mobilise protest and gather political resources. While the Gorleben movement inspired the wider anti-nuclear discourse its final triumph came once the mass protests occurred across the country in the aftermath of Fukushima. Thus local and national cross-cutting coalitions were coordinated and integrated to provide an irresistible power. Throughout most of this long drama, the pro-nuclear forces managed to open and operate the interim store and to ensure the mine was maintained in good order despite the moratorium. The interim store still exists and the mine has not (yet) been closed down and, to that extent, the industry has assets around which it might, one day, regroup. For a long time power relations over Gorleben were shifting but never decisively in one direction or another. But, the eventual victory of the Gorleben movement has not solved the problem of radioactive waste management in Germany; it has shifted it, for the time being, elsewhere or, rather, nowhere.

The Wendland – defence of a peripheral community

To understand why the Gorleben movement has achieved success we need to look closely at the local area and community which was the geographical focus of the conflict. Throughout this book I have used the peripheralisation thesis as a way of explaining the political geography of nuclear conflicts. It is a community-centred thesis providing an explanation of where, how and why conflicts occur. But, peripheralisation is also a dynamic concept offering perceptions and insights into the course and outcomes of conflicts. This is never more so than in the case of Gorleben which displays all the vulnerabilities that are inherent in the periphery but also the processes which can transform the community from weakness into strength, from a position of political impotence to one of political power. Of course, this is not achieved in isolation but within a broader context of changes in the discourses and resources of power in the society at large. But, Gorleben, initially on the periphery, moved eventually to the centre of a broader conflict over the future of nuclear policy in Germany. To explain how this has happened I shall look at each of the peripheral characteristics of the Wendland in turn, beginning with the notion of remoteness.

A remote region

The Wendland is a borderland, an area with fairly clear and identifiable boundaries and cultural features which mark it off from surrounding territory (see Figure 5.4). Its northern boundary is the River Elbe, slow-moving, prone to flooding and wide in its middle reaches. To the south and east lies the border with the former East Germany, an area of country roads, forest and low-lying waterlands which still divides the Wendland from the eastern Land of Saxony Anhalt. It is only on the western side that the Wendland merges rather indistinctly into the wooded agricultural landscapes leading towards Uelzen and Lüneburg.

Within its boundaries the Wendland has a distinctive landscape. Its eastern part, protruding into the former GDR, is known as the *Elbtalaue*, a mixture of waterlands near the river with pastures and abundant bird life, and beyond an area of sandy soils much of it planted with pine and birch forest within which the Gorleben interim store and mine are situated. Although it is poor agricultural land, the heaths and forests provide one of Germany's foremost areas for horse-riding. Moving further west, into the Wendland proper, the soils become more fertile, derived from moraine and loess accumulated during the Ice Age. These deep brown and red soils support farms growing wheat, rape and the potatoes and asparagus for which the area is well known. This is a land of villages and small towns like Dannenberg and Lüchow with a relatively sparse population, around 42 people per square kilometre in the Landkreis and half that in the eastern part compared to 110 ppk² in Germany as a whole. A distinctive feature of the Wendland is the *Rundlings-Dörfer*, the round villages originally developed for subsistence farming and designed for defence in the late Middle Ages. These villages, peculiar to the Wendland, are composed of large, heavy, half-timbered and brick farmhouses (mainly from the

FIGURE 5.12 Wendland cultural landscape

Source: Author

17th to 19th centuries) often forming a ring around a village green with long gardens behind each house bestowing a piecrust formation on the village. Lübeln, the site of an open air Rundlings Museum, is a classic example and around 100 Rundlinge remain. The individual buildings, elaborately decorated and painted in bright colours, originally combined barn and farmhouse. Today they have been converted into large dwellings within the neat, bucolic surroundings which have come to represent the cultural landscape of the Wendland.

This land still conveys a feeling of being cut off from the rest of the country. 'It is in the middle of nowhere in the middle of Germany' as Peter Ward put it (interview, 2014). It is distant from motorways, the nearest well to the north over an hour's drive away (two hours for lorries). There is only one railway link, a branch line from Dannenberg to Lüneburg with few passenger trains a day although this provided the access route for the CASTOR casks. When the Gorleben salt dome was selected the area was on the periphery of the country with the Iron Curtain marking the boundary between East and West. It was, says Lilo Wollny, 'Far away with not much industry and with poor people', seemingly a natural choice for a nuclear plant. Since reunification Gorleben has been in the middle of the country, a kind of internal periphery rather like Bure in France (Chapter 4). It still has this air of remoteness especially to those from elsewhere in Germany. Susan Matthes whom I interviewed in Bonn in 2014 described this combination of political centrality and geographical periphery which characterises the political geography of Gorleben. 'For many years the only place was Gorleben. It was the end of the world'. But, over the years Gorleben has moved more and more to central stage.

Economic backwardness

This, then, is a countryside off the beaten tracks, a rural backwater still predominantly agricultural with service industries and a fledgling tourist trade. Inaccessibility is a deterrent to manufacturing activities though the largest company, SKF in Lüchow, is also in the energy sector, involved in energy efficiency technology. The nuclear industry's contribution to employment, though modest in the region, is significant in the local area around Gorleben and Gartow. The numbers of employees at the mine and at the interim store have dwindled from a peak of around 600 at both facilities in the mid-1990s (400–450 salt dome, 200 in the store and conditioning plant) to around 200 in 2003 (73 and 120 respectively). By 2014 it was around 255 (120 and 135) but declining as the mine closure took effect. The loss of workers with substantial wage packets and the impact on supporting service activities obviously weakens the local economy. The suspension of activities at the mine in 2000 induced economic depression. Gartow's then mayor described the consequences: 'It is an economic disaster. There are no new houses. Flats and houses stand empty and there are no people to inhabit them. Hotels, restaurants and shops have shut down' (2003). In the wake of the brief reprise brought about by the resumption of exploration the declining employment at the mine could leave just 50 workers on a site with an uncertain future and Bories Raapke, its managing director, was pessimistic about its prospects.

This has always been an area of relatively high unemployment, among the highest regional rates in the country with nearly a fifth of the workforce unemployed (Mertens and Haas, 2005). For the local mayors and councils, concerned about economic well-being, the nuclear industry represents a route to greater prosperity. The decline of employment at once both strengthens the argument for the continuing presence of the industry while reducing the numbers of those able and willing to promote the pro-nuclear case. Beyond the councils, any local pro-nuclear movement focused on economic issues has never got off the ground to provide an effective counterweight to the anti-nuclear protesters. Local management has been restrained from commenting, visits to the mine have been discontinued and official comment has been placed in the hands of the BfS (Federal Office of Radiation Protection) whose leadership has been politically ill-disposed towards Gorleben. Promotion of Gorleben locally relies on the workers' committee chaired by Peter Ward. In an economically backward area nuclear activity is no longer a springboard for economic prosperity. In terms of development of the nuclear economy it would appear the power of protest has proved stronger than the power of investment.

Political powerlessness

During the earlier stages of the struggle both sides in the conflict experienced a sense of powerlessness, the feeling that decisions are taken elsewhere and that, despite the struggle, they have little influence over outcomes. In part this was a response to the political structures of power which determine where and how decisions are taken. At the community level local councils (*Gemeinden* and *Landkreis*) have mainly parochial and local regional responsibilities (street maintenance, sewerage, fire services, etc.). They also have responsibility for building regulations and land-use planning which offers potential areas for conflict over proposals. However, the mine was developed under mining law as a research facility which obviated the need for public debate and the surface facilities for both mine and interim store were constructed without any change in the land-use plan. The Land is the authority for implementation and licensing for nuclear waste repositories which come under the Atomic Law. The public must be informed and proposals are subject to public discussion. This process provides opportunities for opponents to challenge a project as was the case with the repository at Schacht Konrad which was delayed for 20 years before the licence was granted in 2002 and, even then, it was held up by individual lawsuits. Although the licence has been granted there remain a few more obstacles, such as transport, before it will become operational sometime after 2020. The experience of Schacht Konrad indicates the scope for obstructing an even more controversial project like Gorleben should an application for a licence ever come to be made. However, it is the Federal Government and its agencies which exercise the greatest power. The federal environment ministry (BMUB) is responsible for policy which is enacted through the Atomic Energy Act. Its agency, the Federal Office for Radiation Protection (BfS), both grants licences to existing operators (for example for transport movements) as well as applications from them to implement new projects such as interim stores or a repository.

Ultimately, the power to decide and implement policy for nuclear waste resides with the Federal Government. Policy has reflected the changing political context especially at federal level. The shifts in policy emphasis create uncertainty which underlines the sense of powerlessness experienced on both sides of the local conflict. For those local councils which have consistently supported the Gorleben projects, the ten year moratorium on investigation at the salt dome created a feeling of immense frustration. 'It's left hanging in the air – we need to know what the actual position is in Gorleben. Is it suitable or not?' (Mayor of Gartow, interview, May 2003). This need for greater certainty was echoed, for opposite reasons, by the protesters. Lilo Wollny, former Green MP, expressed the feeling of dismay thus, 'I've lost a lot of hope. After all these years I thought they would listen, would talk to us, but no …' (interview, September 2003). Yet, she also recognised the ambivalence in the attitudes expressed: 'People wouldn't keep going unless they felt able to influence.' Over the long campaign a sense of powerlessness at the seeming inevitability of defeat was balanced with a sense of hope that things could change, that it was worth carrying on the fight. And that hope was fuelled by the knowledge of victories along the way. If the protesters were powerless there would surely have been no point in persisting in a futile campaign. It was sheer bloody-minded determination that kept them going. As Asta von Oppen expressed it: 'We are not able to lose. We are stubborn' (interview, September 2003).

So, there was as much a sense of surprise, almost of disbelief at the comprehensiveness of the ultimate victory for protest. For Gorleben had become part of the wider anti-nuclear protest which brought about the phase-out of nuclear energy in Germany. And yet, there is still the fear that so long as the mine exists the battle is not over. For the protesters there is pride in achievement but fear that victory is pyrrhic. 'We cannot see the success because we are too involved' said one of the movement's leaders, Wolfgang Ehmke (2014). There is a similar reluctance to let go among the industry's supporters. Realistically they recognise the industry's power has waned dramatically but, like their opponents, they also perceive that so long as the mine is not shut down it could be an option in the site selection process. This may be clutching at straws but it is a better position than having no straws at all. Over the course of the conflict the sense of powerlessness has shifted in tune with changing power relations.

Community and identity in the Wendland

The feelings of geographical isolation, of economic backwardness and of political powerlessness help to define and explain the combination of pride and defensiveness that is characteristic of the anti-nuclear movement. This ambivalence manifests itself in the idea of the Wendland as a special place, but a place that is embattled and fighting for its integrity and even survival. The feelings were summed up by Asta von Oppen during 2003: 'There was identity – we are the last edge of the country. No one wants us, we are supposed to be ignorant and not well liked. We are lacking in self-confidence'. There is present a deeply rooted conservatism redolent of traditional values mediated through integrating institutions of church (still a powerful element in the protests), community and rural lifestyles. It is symptomatic of the power of the movement that it has been able to draw on these traditions and harness them to attack an industry which is the very symbol of modernity. There is an element of David versus Goliath that is evident in the symbolism, slogans, demonstrations and images that accompany the protests. It is through this appeal to the cultural identity of the Wendland that the movement is able to weld together such disparate groups as farmers and city workers, indigenous and adventitious residents, conservatives and socialists, old and young.

FIGURE 5.13 Wendland, landscape of protest

Source: Author

The nuclear presence in the Wendland has precipitated a cultural and political change. In the past, according to Lilo Wollny, this was a 'very remote area, the people very conservative and royalist' (interview, 1995). Nearly two-thirds of the population voted for the Nazis in 1929. In the post-war period the CDU carried around 60% of the vote. It was a region that mistrusted the fragile democracy of the inter-war years but which generally supported the post-war governments. For many people, the coming of the nuclear projects shattered that trust. The industry promised money and progress, 'nuclear power brings heaven on earth' (Eckhard Kruse, Pastor at Gartow, interview, September 2003). Over time, suspicions were aroused as it became apparent that Gorleben had been selected as the site for the perpetual management of Germany's wastes. There was a feeling that the choice of Gorleben was unjustified and unfair. Attitudes were transformed. As Pastor Kruse put it, 'In Gorleben there was no one who dared say no. Now there were X's all over the place' (reference to the yellow crosses, the ubiquitous symbol of protest in the area). Of course, the majority of the population here, as anywhere, remained passive or indifferent to the issue. But, for those who became engaged, 1977–9 was a turning point, a time when a hitherto conservative area became radicalised.

It was from this moment of transition that the idea of the Wendland as a cultural identity was formed. The Wends were a Slavic tribe who settled in this area during the late Middle Ages, part of the movement of peoples back and forth across the north German Plain (Christiansen, 1998, see quote at head of chapter). They possessed their own traditions and language but these gradually died out so that the Wends became an obscure folk memory. The Wendland, to which the Wends gave their name, was rediscovered, or rather reinvented, by the anti-nuclear movement. Asta von Oppen recollects there was something called Wendland but only in museums – 'it was more of a joke'. Wendland was called into being and gradually it came to represent the whole area of Lüchow-Dannenberg. The protest movement invented its flag (a bright orange pointed sun on a green background) and created the idea of the Free Republic, independent minded and free of the nuclear industry. Soon the concept of Wendland was taken up by the tourist industry and promoted as somewhere to seek 'peace and seclusion and pure nature' (tourist leaflet). Its wider cultural significance was proclaimed in the application of The Rundling Cultural Landscape as a UNESCO World Heritage Site which provoked the remark made to me that 'The UNESCO site would not be the rundlings but the CASTOR protests' (Hans-Udo Maury, local politician, 2014). The cultural notion of the Wendland emerged to become a powerful central idea, a focus for the anti-nuclear movement. Its integrity had to be protected against the violation of the nuclear industry. Here, under an external threat, history and geography have been called upon to provide a unique territorial identity around which to mobilise the forces of resistance.

Environmental risk

Gorleben is a nuclear oasis that is only half formed, reflecting its history. Gorleben is not exactly a greenfield location but neither is it a fully established nuclear reservation. The nuclear industry has a foothold, to be sure, established but no longer operating in the case of the store, and excavated but moribund now in the case of the mine. With the cessation of CASTOR transports the environmental risks have reduced and are now confined to the store itself which is heavily protected and shielded. It was the prospective risks to present and future generations from permanent deep disposal in a repository that was the main motivation for the Gorleben protests manifested in actions against the transports and at the mine itself. The risks from radioactivity that might have been concentrated at Gorleben are presently dispersed in stores across Germany and in France and the UK and will continue to be so until a site for a repository is found.

The Wendland was for long an area in transition as traditional rural values were exposed to the forces of modernity represented by the invading nuclear industry. From its peripheral position the region became central to the struggle over nuclear energy. Economically backward it has nonetheless resisted the power of investment, jobs and economic regeneration the industry represents. Politically, once powerless, it has nurtured a protest movement comprised of a loose but effective coalition of conservative and radical groups, determined to protect the cultural identity of the Wendland, an idea invented to emphasise the uniqueness of the area. And it has linked in to a wider protest against nuclear energy drawing in an ultimately irresistible power capable of routing the nuclear energy industry itself. For over three decades the struggle continued against a background of shifting political forces favouring first one side then another until a decisive victory was in sight. Throughout that time there was continuing uncertainty over the long-term management of Germany's nuclear wastes. Gorleben, though a potential option, was initially an experimental project, an exploratory salt mine developed for the conduct of experiments, measurements, assessment and evaluation of containment within a salt host rock formation. During all that time various efforts were made to seek ways out of the impasse of German nuclear waste policy.

Nuclear waste policy in Germany – the search for consensus

Throughout its history German nuclear policy including nuclear waste has resulted in polarisation of views and confrontation. Gorleben is the archetypal struggle. From time to time efforts have been made to cut across the divisions, to assemble the warring factions, to encourage dialogue and consensus. The struggle was between two parallel discourses. One was the dominant discourse of Danger and Distrust from which the anti-nuclear movement derived its power. The other was the discourse of Consensus and Cooperation, subordinate but occasionally effective, as in the consensus on nuclear phase-out combined with an effort to find a process for the long-term management of nuclear wastes at the turn of the century and renewed a decade later. A consensual process would have to grapple with dealing with the inherited and inherent conflict over Gorleben.

The first initiative was taken during the early 1990s when consensus talks were held between the Federal Government, the Länder governments, the trade unions, the utilities and nuclear industry and environmental groups. Although there was common ground on such matters as energy conservation and the use of renewables, the talks ultimately served to confirm the fault lines on policy between pro- and anti-nuclear interests. As the new century dawned, a further major initiative was taken to set out a comprehensive and consensual approach to long-term radioactive waste management.

The AkEnd process

The political circumstances for such an approach were propitious. At the federal level a Red/Green coalition led by Federal Chancellor Gerhard Schröder, former Prime Minister of Lower Saxony, agreed to a complete review of nuclear policy as a *quid pro quo* for Green participation in the government (Rüdig, 2000). Accordingly, in 1999 the Green Minister for the Environment, Jürgen Trittin, established a Committee on a Site Selection Procedure for Repository Sites (known as AkEnd from its German title). AkEnd comprised 14 experts including natural scientists, regulatory bodies and the nuclear industry but also a social scientist and a member from the Öko-Institut, a prominent and influential environmental research group. Its job was 'to develop a procedure and criteria for the identification and selection of the best possible site for the safe disposal of all types of radioactive waste' (AkEnd, 2002a, p.1). During its working period of three years AkEnd held public workshops, met with a wide range of organisations in government, industry and civil society and through meetings, publications and dialogue tried to understand the views of the public on the issues of radioactive waste management. It produced its findings and recommendations at the end of 2002 in a comprehensive report which sought to identify how to achieve societal consensus on the long-term management of Germany's radioactive waste (AkEnd, 2002a, 2002b).

The committee had freedom to reach recommendations on siting within certain basic constraints. The first of these was the legal requirement to focus on deep geological disposal as the only option for long-term management, on the grounds that 'there is no long-term safe and ethically justifiable alternative to the disposal of radioactive waste in deep geological formations' (2002a. p.6). Thus, putative solutions or variants being actively explored in other countries (for example, France, Chapter 4) such as long-term storage, retrievable disposal or partitioning and transmutation were rejected. The committee affirmed its view that these solutions all involved keeping options open and so passed some of the problems of managing wastes to future generations. 'This implies a very far-reaching requirement regarding sustained stability of today's social system with its ethical values and standards for a long period of time. Experience shows a substantiated prediction on this issue cannot be made' (2002b, p.24).

Although required to limit its considerations to deep disposal, the committee was firmly against the idea of retrievability which was being considered in other countries (again, France is a good example) as the best way of gaining public acceptability and enabling future generations to have some involvement in decision making (Kommentus, 2001; Hinchliffe and Blowers, 2003; Aparacio, 2010). It reasoned that the need for continued monitoring would impose burdens of cost and risk on future generations and prevent the point at which the wastes could be regarded as in a condition of passive safety. As Detlev Appel, a member of AkEnd, observed during an interview with me in 2003, 'What are you waiting for in the context of the time frame of radioactive decay?' There was also the possibility that the geological conditions which favoured retrievability might not be the most favourable with regard to long-term safety. For instance, one of the safety features of salt formations is claimed to be that salt encroaches on the waste making retrievability difficult.

A second constraint on the committee's work was that all of Germany's radioactive waste would be placed in a single repository. While this one-site solution would potentially reduce the problem of finding alternative sites, it raised both technical and political issues. Technically, the ability to accommodate all wastes within one repository had scarcely been explored. Politically, it was not feasible since the deep abandoned iron ore mine Schacht Konrad, also in Lower Saxony, was under consideration for the disposal of non-heat generating wastes.[2] In practice, as Klaus Kühn, an AkEnd member observed, the committee believed it was looking for a site that would take the heat generating wastes (interview, 2003).

Thirdly, the government recognised it was 'obliged to promote the disposal of radioactive waste on German territory'. As a substantial waste producer Germany was here merely adhering to the principle of national self-sufficiency which inhibits the international trade or transfer of radioactive wastes. This was in line with the IAEA's Code of Practice which says that 'it is the sovereign right of every state to prohibit the movement of radioactive waste into, from or through its territory' (HMSO, 1995, para. 142). With the ending of reprocessing overseas, Germany would be able to achieve self-sufficiency in the management of its wastes. However, considerable volumes of German waste residues still awaited repatriation from France and the UK and, until the contracts were fulfilled, the transport of CASTOR casks to Gorleben would continue to fuel the conflict over radioactive waste.

A fourth requirement was that site selection should not exclude any area in advance. 'The project starts with a "white map" of Germany. All areas are dealt with equally' (AkEnd, 2002b, p.65). This meant that salt was no longer the preferred host rock for the repository; the aim was to find the most favourable overall geological setting. Selection could then proceed through the application of both geo-scientific and socio-scientific criteria. If possible five sites (but no less than three) would be identified for surface exploration. Ultimately, two sites would be identified for detailed investigation. In principle this process appealed as both fair and rational but the 'white map' contained one spot, Gorleben, where investment had already been made into an exploratory mine. The nearby storage and conditioning facilities represented further investments into Gorleben as the final destination for high level wastes. If somewhere else were chosen, the question arose as to who would pay. The utilities had already sunk investment into a facility where, so far as they understood, nothing had been discovered which would rule it out on technical grounds. So there Gorleben remained, a potential candidate, both asset and obstacle in the search for a site.

A fifth requirement was that the choice of sites for underground exploration had to be completed by 2010 in order that the repository could be ready for operation in 2030. The AkEnd Committee regarded these targets as 'very ambitious in view of the tasks to be coped with within this period' (2002b, p.19). In particular the 2010 deadline would require 'the rapid legitimisation and implementation of the selection procedure' (ibid.). In view of the previous history of conflict over radioactive waste management, the deadlines might justifiably have been described as heroic.

New approaches to resolution

It is worth looking at the AkEnd report as an innovative and imaginative approach to the problem of siting a geological repository, comparable in a number of ways to the slightly later CoRWM report in the UK (see Chapter 4). The process recommended by AkEnd envisaged three phases: the first, completed in 2002, was defining the step-by-step procedure for site selection; the second was getting political and societal agreement to the procedure; and the third was the implementation phase leading to the siting and development of a repository. During the first phase AkEnd set the criteria to be used in the assessment of suitability of sites and the decision making process that would be followed.

A fundamental requirement was that wastes in the repository should be isolated from the environment for a period 'within a magnitude of one million years'. Geo-scientific criteria were paramount in ensuring the integrity of the repository. Geological areas that would be excluded from selection were those where there were large vertical movements, active fault zones, seismic or volcanic activity or where groundwater flows were unfavourable. Minimum requirements in terms of depth, extent and thickness of the host rock were set out as well as the requirements for a favourable overall geological setting which included slow water movement, suitable rock type, good retention capacity and so on. The specification of geological properties, derivation of criteria and the assessment methodology were elaborated in some detail in the AkEnd report underlining the concern for safety.

Although it might be said that the approach was geology-led, the Committee also concluded that scientific criteria alone were not sufficient and that 'it will only be possible to realise a repository site with success if socio-scientific aspects are taken into account' (AkEnd, 2002b, p.178). Some of these aspects are reasonably straightforward, notably the need to avoid nature reserves, natural parks and other designated areas, to protect valuable forested and agricultural areas and to keep reasonable distance from mining, urban development and major infrastructures. More problematic was defining the development potential of site areas in terms of whether a repository would prove beneficial or detrimental. Here the Report advocates 'potential analysis' combining both 'mental and material' aspects of development to be assessed by both citizens and developer of the repository.

This idea of development potential is related to the concept of 'perspective compensation' which was another novel aspect of the proposed selection process. The idea is that citizens in areas of potential sites identify their vision of the future for their region. For example, a rural region might envisage its future in terms of tourist potential, an industrial area in terms of the development of environmental technology. 'The regional development concept must emerge from within and be realised with the initiative of the region' (Ipsen, 2003, p.6). Funding would be provided to those forms of self-initiative necessary to realise the concept. The idea was to provide opportunities for the long term based on self-determination. It was based on the concept of 'region' which is recognised as 'home' by 80% of the population (AkEnd, 2002b, p.205). Regions would emerge through agreement based on citizen participation. The late Detlev Ipsen, a sociologist and AkEnd member who was mainly responsible for the idea, argued, 'If regional building is a process then it cannot be determined in advance'. He described it as 'an integrated sociological concept' (Ipsen, 2003, p.7). When I met him in 2003 he presented his ideas as 'a combination of vision and volunteering' designed to realise regional potential in the area where the repository is eventually built. It would provide both incentive and compensation but only on the basis of participation by those communities which have consented to host a repository.

The regional development concept emphasises citizen participation. Indeed, participation is integral to the whole process of site selection. AkEnd perceptively recognised that 'Participation does not guarantee success in finding a repository site, but a lack of participation increases the chances of failure' (2002b, p.58). Surveys of public opinion had indicated a lack of trust in institutions, a demand to be informed and involved in decisions, a desire to find a solution as soon as possible and a preference for solutions to be voluntary rather than imposed (2002a, pp.14–15). Consequently, there was a need to build trust by engaging in dialogue, through an accessible and transparent process, by enabling citizens to achieve the competence necessary to participate and by giving them control and responsibility for decisions. But, as in every country, Germany was also faced with the paradox that while the bulk of the population regards a solution as urgent, 80% reject the idea of a repository in their region (2002b, p.210, fig. 6.3).

This problem was tackled by setting out a step-by-step procedure of participation for the implementation phase. During each step citizens would be provided with adequate information and support from experts. The first two steps would evaluate the criteria to identify areas with favourable geology. Step 3 would identify and select site regions for exploration from the surface. The potential of regions would be explored and citizens would indicate willingness to participate by a vote to be binding on the local councils. On the basis of the assessment of the sites, at least two would be chosen for underground exploration. Again, citizens would control the process by a vote whether to proceed which must be ratified by the councils. The final step was an assessment of the two sites and an 'orienting' vote of citizens and councils which would be used to help guide the Bundestag to make the final decision.

This procedure provides both citizens and their political representatives with considerable control over decision making. They have the power of veto right up until the final stage where their views are considered though not binding on the final decision which is Parliament's. In Detlev Appel's view this approach was 'quite revolutionary in a German context' (interview, May 2003) where democratic institutions tend to be rather inflexible and rule-bound. It represented what his colleague AkEnd member Klaus Kühn described as an exercise in 'basic democracy' encouraging widespread debate and participation on the part of civil society. But, Kühn warned, 'If you follow our process then Parliament is not involved – so where is their role?' (interview, May 2003). According to Detlev Ipsen the participatory elements are complementary. 'Representative democracy has to take the final decision' (interview, May 2003). The answer seems to be that Parliament steps in if all else fails. AkEnd declared, 'the civil self-organisation is not only an alternative to the representative democracy, but is only politically effective through and in reference to it' (2002b, p.53). The combination of self-determination, regional development and control of the process encouraged AkEnd to believe that its approach would ensure that 'the identification and selection of a site can in general be carried out and concluded successfully' (ibid., p.204). But, it also recognised that, given such devolution of power, it is conceivable that there would be too few volunteers at the various stages, in which case the Bundestag would be left to decide how to proceed.

The AkEnd report was far-sighted, reflecting the latest thinking on how to go about finding a host community for such an unwanted project as a deep geological repository. From a policy of one site at Gorleben in salt the issue had been opened up to one site anywhere in any feasible rock formation. Moreover, the selection of the site and host rock would be based on an unprecedented exposure to public debate and local control over decision making based on volunteerism and veto. It was an important statement of the need to integrate the scientific and social aspects of site selection. It set out detailed criteria both scientific and social that would have to be applied in identifying and choosing a suitable site. It emphasised the need for phased (staged) decision making and for comparative assessment. It recognised the central importance of citizen participation and consent. It distinguished between participative democracy to achieve public support and representative democracy to ensure legitimation. And, it introduced concepts of regional development, well-being and compensation deemed essential to secure the willingness of a community to participate. Although it did not, in the end, proceed in Germany the AkEnd report stands as a truly original approach to tackling one of society's most intractable problems. And, it also had some influence on subsequent thinking elsewhere, not the least on my own thinking in drawing up the recommendations for implementation of CoRWM's proposals in the UK (CoRWM, 2006, 2007). Those recommendations, at least, were taken up and formed the basis for the UK's approach to site selection for its GDF (see Chapter 3).

From impasse to breakthrough – Germany's anti-nuclear moment

Gorleben – an uncertain future

The AkEnd process never got beyond its first stage, setting out the approach to site selection; the next stage of the process, the so-called Phase 2, discussion and agreement on the procedures detailed in the Report, soon ran into the sand. Some key interests were reluctant to participate in negotiations: the nuclear utilities were unwilling to fund alternative proposals given they had already funded the Gorleben project; the Christian Democrats and their partners, the Free Democrats, were unsympathetic to proposals emanating from the Red/Green Government; the Länder were concerned to avoid hosting a repository; and some environmental groups, notably Greenpeace, were sceptical about the real purpose of the process. In any event, the issue had lost political momentum with the SPD noticeably less enthusiastic than their coalition partners, the Greens in the Federal Government. As Bruno Thomauske, a member of AkEnd and representative of BfS, told me at the time, 'This is not an important question at the moment. There is no great pressure' (interview, May 2003), a view echoed by other colleagues like Michael Sailer of the Öko-Institut. Years later, the disappearance of AkEnd still rankled among those who had been involved in the process. Jörg Mönig of GRS who had drafted the final AkEnd Report expressed to me in 2014 his disappointment that it was not taken up. 'It just vanished – there was no political will. There was some public impetus but no one wanted to take it up'.

For a few years the AkEnd process had offered some hope that the impasse could be ended. As Rebecca Harms put it, 'In Gorleben the process was interrupted and it seemed there could be a more scientific process. But, now hope is disappearing' (interview, October 2003). With the AkEnd process in abeyance the prospects for alternative solutions diminished for a while. The idea that volunteers might come forward remained untested in the German context. Bruno Thomauske was quite candid on the issue of volunteers. 'I'm pessimistic as to whether we will find any candidate who would propose itself' (interview, May 2003). This scepticism was shared by Susanna Ochse of Greenpeace. She considered AkEnd was 'not in the real world' where 'the reality is having sites already' (interview, October 2003). But in not commenting on the situation and suitability of Gorleben the Committee had failed to remove the source of conflict.

In any event, the years following the AkEnd report were a time of stasis in the nuclear debate. Unlike in the UK there was no revival of political momentum for new nuclear power, rather the reverse as phase-out, whether rapid or protracted, was widely accepted. As far as waste was concerned there was some progress notably at Schacht Konrad and interim storage of spent fuel and high level wastes was the approved means of management for the foreseeable future. Even at Gorleben the eventual lifting of the moratorium on exploration of the mine in 2010 represented a commitment to continuing research rather than any more active development though it reaffirmed its status as the only potential site for a deep repository for highly active wastes.

> So one could lean back and leave it up to the experts to test the proposed site(s) and to the local and national opposition to make sure that the process was slow enough to prevent any real progress. Anybody seemed to benefit from paralysis.
>
> (Hocke and Renn, 2009, p.931)

That is, until Fukushima reanimated the anti-nuclear movement across Germany and swept away any possibility of a nuclear revival. The post-nuclear age of *Energiewende* 'Energy Transition' had begun.

It is important to recognise that anti-nuclear action had intensified before Fukushima, in the previous year, 2010, as a response to the Centre-Right Federal Government's proposals for slowing down the nuclear phase-out (agreed by the previous Red–Green Coalition to be achieved by 2021) by extending the lives of some of the nuclear stations. This precipitated a mass demonstration in April in the form of a human chain from the nuclear plants of Brunsbüttel on the Elbe estuary passing right through Hamburg to the Krümmel plant further up the river, a distance of 120km and involving 120,000 people. There were simultaneously demonstrations at Biblis power station (on the Rhine near Mannheim) and Gronau (in the northwest on the Dutch border). Later that year, at the time of the November demonstration against the shipments to Gorleben, an estimated 50,000 demonstrators came to the Wendland to protest against nuclear energy. On the very eve of Fukushima a human chain of 60km stretched from the Neckarwestheim power station to stage a rally in Stuttgart as a protest against the Baden-Württemberg Land Government's pro-nuclear stance. In the immediate aftermath of Fukushima there were protests all round the country with four major demonstrations in Hamburg, Munich, Berlin and Cologne involving an estimated half million people. As Anika Limbach of AntiAtomBonn commented, 'In Germany never before and afterwards had there been mass demonstrations of this dimension' (personal communication, 3 April 2014).

Whatever the numbers involved (and estimates are always contested) both in scale and composition (including a broad spectrum of society, churches, mainstream environmental groups and major political parties, SPD and Greens) such a spectacular public manifestation could not be politically ignored. And, indeed, it was not. Public opinion, already heavily against building new nuclear power stations became almost universally so after Fukushima, moving, according to one poll, from nearly three-quarters against in 2005 to 90% in 2011 (BBC, 2011).[3] The Federal Chancellor, Angela Merkel, duly noted the political weather and in a *volte-face*, remarkable in its combination of political courage and pragmatism, announced, in May 2011, the phase-out of nuclear energy by 2022 with eight stations to close immediately, coupled with a commitment to renewable energy and energy efficiency. The decision to end nuclear energy was informed and justified by a report from a specially convened Ethics Commission on a Safe Energy Supply chaired by Klaus Töpfer, a former Federal Minister of the BMU (Ministry for Environment, Nature Conservation and Nuclear Safety) and including among its members such eminent social scientists as the late Ulrich Beck known for his work on *Risk Society* (1992) and Ortwin Renn, a leading exponent of processes of public deliberation and dialogue (2008). The Commission stated that Germany 'must direct its attention towards an energy supply that dispenses with nuclear power as quickly as possible and that promotes Germany's path towards a sustainable development and new models of prosperity' (Ethics Commission, 2011). The Report's central idea was the avoidance of risk of limitless consequences to environment and people that Fukushima had demonstrated. 'Accordingly, the conclusion drawn is that, if adverse events are to be ruled out, nuclear technology must no longer be used.' Moreover, the ethical position was supported by the fact that 'Germany is able to replace nuclear power with lower-risk technologies in an ecologically, economically and socially acceptable way'. Nuclear energy had 'poisoned atmosphere in society at large' and, therefore, the ethical way forward was nuclear free.

To an extent the issue of nuclear wastes seems peripheral to the ethical and political tide that overwhelmed the German nuclear industry after Fukushima. Yet, the turn around of 2011 also unlocked the impasse on radioactive waste management policy that had set in after the AkEnd report was shelved. AkEnd had been a missed opportunity (Hocke and Renn, 2009, p.933) but, so long as Gorleben was the only site in the frame, it was likely to be fiercely resisted. The Ethics Commission entertained the idea of retrievable storage which 'expands the area in which final sites for radioactive waste can be sought in Germany beyond Gorleben' (2011, p.40). And, with nuclear energy on the way out, a solution to the problem of managing wastes need no longer be ethically compromised by the possibility of legitimating new nuclear energy. So, in a renewed effort at consensus brought about by the changed circumstances post-Fukushima, the way had become clear for another attempt to find a solution; but would it be with or without Gorleben?

Another new beginning

One of the outcomes of the consensual transformation of energy policy in Germany was the reanimation of the effort to find a permanent solution for the management of highly active radioactive wastes. By 2012 spent fuel assemblies and highly active wastes from reprocessing in 984 casks were scattered around the country in interim stores at 12 reactor sites, former central stores at Gorleben and Ahaus, the research centre at Jülich and in a store at the former Greifswald nuclear site (Völzke, 2014). There were an estimated further 500 casks needed to accommodate all the wastes arising by the time of final shut-down in 2022. The decision to stop further shipments of repatriated wastes from the UK and France into the Gorleben interim store was a necessary *quid pro quo* to secure the support of anti-nuclear groups for a consensual approach to waste management. Licences on the interim stores were limited to 40 years and all would fall in well before mid-century though extensions were possible. In the context of nuclear phase-out, the ultimate size of the inventory was known, the need for a repository was agreed and there seemed a political willingness to get things moving.

Accordingly, the Federal Site Selection Act on the Storage and Disposal of High Active Waste and Spent Fuel in Germany came into force in July 2013. Transparency and participation of all citizens in all procedural steps was 'a prerequisite for a decision that is supported by a broad consensus'. The Act established a Commission to develop the criteria and procedure for site selection to report during 2016. The mandate for the Commission was very broad indeed. The Commission was to begin its work with a clean slate, a 'white map' with no identified sites or areas, to cover alternatives to deep geological disposal, and to set out requirements for public involvement. The one site (Gorleben), one rock type (salt) approach that had been the ineluctable basis for policy for nearly 40 years was swept away by a new broom with all options open. The balanced membership of the Commission consisted of a Chair and 32 members divided into four sectors comprising eight each from the Federal Parliament, the Länder, science and civil society (respectively two each from industry, religion, environmental groups and trade unions). Although it was intended to put together all the elements necessary to achieve a consensus the membership clearly contained the ingredients for conflict. This became immediately apparent with the reluctance of some environmental groups, including Greenpeace, to participate.

There was a degree of relief that a new beginning was in prospect though mixed with scepticism that any real progress could be made within the short timetable outlined. Jörg Mönig of the research group GRS was happy the law was passed since 'for the first time there was broad support' (interview, 2014). His hope was that 'phase-out would make it easier to come to terms with finding a solution'. Georg Arens, a senior official with the environment ministry BMUB and formerly with AkEnd, was doubtful that there was sufficient political commitment.

> In my opinion no one is really interested in finding a new site. There is not really strong support from politicians to find a new site. This law won't avoid new resistance. It reflects only a moment in time – consensus has to be maintained for 10, 20, 30 years.

According to Arens the reality was that waste management was safe and well organised and the search for a long-term solution was unreal; it carried neither political bonus nor urgency. 'It is a theatre' (interview, 2014). On the other hand, Michael Sailer believed the search for a repository site needed to be taken seriously. While the storage casks could last for up to 60 years there would eventually be a problem of degrading fuel and declining safety.

> The reality is that if you don't solve it you will have a problem at all interim stores. The public don't realise this is a real threat. We are looking at a process

that takes a long time. There are crazy debates because we need a process which has a possibility of a site and a facility by the 2040s. In a technical sense it is possible. In a political sense it is not so.

(Interview, 2014)

There is another reason why a consensual solution may not easily be found – in a word, Gorleben. It seems impossible to make progress with Gorleben in the mix but also impossible without it. The site selection act begins with a *tabula rasa*; it rules no site in but, at the same time, none is ruled out. No further work on Gorleben is to be undertaken, yet so long as the exploratory mine exists it must be maintained. Michael Sailer explained the dilemma, 'You cannot exclude Gorleben and it is not excluded but it is not included as a preferred location. Everything must be done to give no preference. There is no scientific argument against it yet. So, if it is excluded the reason would be political'. But, as Jörg Mönig recognised, a successful policy may only be possible if Gorleben is taken out – 'it is politically burnt'. Paradoxically, it must be kept in for political reasons.

> My feeling is that if you take a site out for political reasons then you can't convince the public at another site since it has not been compared. It is a question of fairness for me. If you follow the process you can't simply take one out at the beginning.

Gorleben persists as a divisive issue. However, on one thing both sides seem to agree, that nothing much will come of the new process since, whether or not Gorleben stays in, nowhere else is likely to come forward in the near future. Speaking to me in 2014, Bories Raapke, Managing Director at Gorleben, was categorical, repeating the point that, 'In the process of site selection it may be that the decision falls back to Gorleben one day'. Funding will be a problem and the nuclear industry has already sunk resources into Gorleben and, in its declining state, will hardly be enthusiastic about underwriting a site elsewhere. As Georg Arens saw it: 'Site selection will be funded by operators but all the time Gorleben is still there. Gorleben is not officially given up but everyone recognises the low probability that Gorleben will be realised' (interview, 2014). It appears that the long struggle over Gorleben has resulted in stalemate. Despite the new beginning promised by the Site Selection Act, Germany, as elsewhere, seems in for a long and arduous search for an acceptable and viable solution for the long-term management of its spent fuel and high active wastes.

The power of persistence

The key features of the Gorleben protest movement have been its persistence, its ability to respond to changing circumstances and its success in building a powerful and cross-cutting coalition able to mobilise support from within the local community as well as from the wider anti-nuclear movement in Germany and beyond. In common with both local and national anti-nuclear movements in Germany, the Gorleben movement has 'maintained its radical and uncompromising stance' (Rucht and Roose, 1999, p.72). And, over its long history, although it has suffered setbacks it has, overall, achieved success.

The Gorleben movement has both flourished and survived during a time of shifting nuclear discourses and consequent changing power relations. When it all began during the 1970s, the power of the nuclear industry backed by sympathetic governments at Länder and federal level may, objectively, have seemed indomitable. After all, the Gorleben salt mine had been plucked out of the ground, so to speak, as the site for the burial of the nation's accumulating highly active radioactive wastes. Along with plans for a reprocessing works, interim store and encapsulation facility, the mine was an integral part of the *Entsorgungskonzept* imposed as a *fait accompli* by government on an unsuspecting community. Such Decide Announce Defend approaches were typical outcomes of a discourse which was both pro-nuclear and undemocratic. But, this was also a time of growing environmental awareness with the development of both mainstream environmental movements and local initiatives as well as more radical formations (Rucht and Roose, 1999; Doherty, 2002). The Green Party was also emerging at this time. The anti-nuclear movement drew on the conservationist, local and radical elements to refine its repertoire of protest and develop an alternative discourse, not only anti-nuclear but embracing ideas of democratic participation, environmental protection and protest. By its commitment and ability to mobilise people across political, geographical and social boundaries the anti-nuclear movement, and the Gorleben movement in particular, was able to exert influence and, at crucial moments, considerable political leverage. Gorleben was and is a distinctive movement but it is also part of a wider movement. In practice, the anti-nuclear movement is a constellation of localised groups, some, like Gorleben and AntiAtomBonn, originating in the Bürgeriniativen scattered across the country. At the national level there were bodies such as Greenpeace, BUND or Robin Wood and professional research institutes which provided expert critiques, of which the most prominent has been the Öko-Institut. The variety of interests and organisations that could be drawn together gave the anti-nuclear movement its dynamism, flexibility and breadth of support in society at large.

At the outset in the late 1970s the anti-nuclear protests focused around Gorleben achieved success. There followed two decades where protest, though continuing and consistent, yielded little by way of palpable outcome. This reflected a period of relative quiescence on the nuclear front generally when at Gorleben the nuclear industry seemed to be consolidating its hold by developing its research at the mine and opening the interim store. It should be said that this was an era before and after reunification, when other issues (employment, social welfare, the economy) were gaining higher priority, to some extent at the expense of environmental problems. Nonetheless, anti-nuclear feeling remained strong especially at local level as reflected in the opposition of people to the possible development of nuclear reactors in their neighbourhood which by the 1980s accounted for over half the population with only about a fifth in favour (Dominick, 1992, p.218).

The power relations which had stabilised in the 1980s began to shift during the 1990s when the protests at the movement of flasks to Gorleben began in earnest. By the turn of the century the phase-out of nuclear energy instituted by the Red/ Green Federal Government precipitated a new initiative on radioactive waste management (AkEnd) and work at the mine was suspended. A revival in nuclear energy's fortunes a decade later, including resumption of research at the mine, was summarily reversed by Fukushima and the *Energiewende* that followed. Once again a new initiative to find a site had begun though, this time, Gorleben appeared an unlikely candidate. The closure of the interim store to further cask shipments signalled that all the elements of the original *Entsorgungskonzept* were moribund. Power appeared to have shifted, this time irreversibly. The anti-nuclear discourse fuelled by the power of protest, for so long focused on Gorleben, had seemingly triumphed.

As at the beginning so, too, at the end success was brought about by a broader, wider national anti-nuclear energy momentum. National groups such as Greenpeace and *Ausgestrahlt* (since 2008 a leading anti-nuclear group) have provided the arguments, organisation and coordination necessary for raising awareness and mobilising the resources of mass protest. But, success has only been made possible by organisation on the ground. As Brand observes, 'It is only the smaller groups that, in part, keep up the traditional style of grass-roots mobilisation, above all among anti-nuclear groups' (1999, p.50). Heinz Smital of Greenpeace expressed to me the crucial role of the Gorleben movement in the process, 'The environmental movement in Gorleben did the important work of integration. It got bigger and bigger and the spirit of protest developed' (interview, 2014).

Ultimately, we can draw on the thesis of peripheralisation to explain the success of the Gorleben movement. It is a demonstration of how a remote, apparently powerless and disadvantaged area can create and reflect a changing discourse and mobilise the resources to defend its cultural and territorial integrity. Gorleben provides the acme of environmental activism, a group characterised by a fixed and unflinching purpose and a rootedness in the local community. The Gorleben activists combined the wider purpose of stopping nuclear energy with the specific determination to prevent the Wendland from becoming the final resting place for the country's nuclear waste. Part of the motivation, at least in the early years, was ideological, a determination to avoid the mistakes of the generation that allowed fascism to flourish. As Marianne Fritzen put it, 'We don't want to be asked by our children, what did you do against nuclear power?'. There was, too, resentment kindled by the lack of justification for singling out Gorleben. There was a single-mindedness about the movement. The Gorleben movement was built upon foundations of friendship, as Lilo Wollny expressed it 'a background of belonging together, sometimes feeling like love for each other – it's difficult to explain'. This closeness, feelings of common identity and community solidified the campaign. The Gorleben movement has maintained its radical, grass-roots characteristics. In some ways, Gorleben expresses an 'old style while the world has moved on' (Michael Sailer, interview, 2003). Perhaps, as Rebecca Harms conceded back in

2003, the movement was too idealistic to recognise what had been achieved. There was a feeling then that it clung on to old oppositional strategies when new approaches were needed to fit changing conditions. And yet, it seems the very persistence of such single-minded purpose coupled with the ability to embrace a broader agenda and capitalise on the weakening power of the nuclear industry has been instrumental in the achievement.

With its central objectives accomplished it may appear that its work is done. During the decades since the movement began, the power relations in Germany have utterly changed. The nuclear industry, long in retreat, will soon be gone although its legacy of waste and decommissioning will remain for decades to come. The Gorleben movement has found it difficult to come to terms with the scale of the changes it has, in part, brought about. The history of conflict and of mistrust between the anti-nuclear movement, government and industry provides the oxygen for continuing protest. The Gorleben movement continues to recruit, motivate and mobilise and there are reasons for it to persist in its activities. One is the fact that a perceived threat continues to exist. The interim store still has capacity and for the moment the pilot conditioning plant is ready for operation. And the mine though inactive is not yet irretrievably closed. Its continuing existence makes protesters, some of whom have given a lifetime to opposing the mine, understandably nervous that it could come back yet again to haunt them. Their long held sense of isolation fuels the fear that a repository is unlikely to be acceptable anywhere else.

Perhaps the most compelling reason for continuing derives from the social dimension of peripherality, that shared sense of identity, of long-standing comradeship and common purpose deeply embedded in the older generation but, to an extent, passed down the generations of protesters. The years of opposing the transports and the repository became, for many people, a deeply ingrained mission, one that they could neither relinquish nor compromise upon. They were motivated by a sense of unfairness, of revolt against the intrusion of a dangerous technology and a belief in preserving the identity of the Wendland for themselves and for the generations to come. Their strength lies in the cohesion of radical and deeply conservative ideologies.

For Gorleben is not an evanescent phenomenon, here today and gone when its purpose fades. While these conditions remain and as long as the nuclear presence remains the Gorleben anti-nuclear movement is likely to maintain its vigilance. Its very vulnerability in the past was a source of determination to resist; its success has opened up the possibilities for dealing with the nation's nuclear legacy without Gorleben.

Visits and interviews

This chapter is partly informed by the several visits I made and the people I met over nearly three decades. Although I visited several nuclear sites in Germny, it will be clear from the narrative that the Gorleben story was endlessly fascinating to me. I have listed most of those I met below and am grateful to all of them.

1986 October. Braunschweig, Konrad radioactive waste facility: County Councils Coalition

I participated as a county councillor on a fact-finding tour of three countries (Sweden, France and West Germany) undertaken by members and officers of three county councils (Bedfordshire, Lincolnshire and Humberside) then engaged in opposing plans for radioactive waste disposal in their areas. The West German part of the visit focused on Schacht Konrad, a former deep iron ore mine where Germany's intermediate level wastes are to be buried. The visit is described in *The Disposal of Radioactive Waste in Sweden, West Germany and France* (Prepared for the County Councils Coalition by Environmental Resources Ltd., January, 1987).

1994 October. Bonn, Hamburg and nuclear waste facilities at Jülich, Morsleben, Konrad and Gorleben, Radioactive Waste Management Advisory Committee (RWMAC)

During a study tour as member of RWMAC I met with government officials, radioactive waste organisations, regulators and Greenpeace together with tours of facilities. The itinerary and programme are in RWMAC Fifteenth Annual Report, May 1995 Annex 5.

1995 May, visit to the Wendland

I observed a protest demonstration at Dannenberg.
　　Interviewed: Rebecca Harms, Green Party Lower Saxony Parliament; Marianne Fritzen, campaigner; Lilo Wollny, former Green MEP; R. König, BLG, CEO Interim Fuel Store; Dr. Rolf Meyer, DBE, Gorleben mine; Mayors of Gartow, Gorleben and Gartow Samtgemeinde (combined communities); Christian Meyer and Herman Müller, journalists on the regional newspaper *Elbe-Jetzel Zeitung*.

2003 May, visit to Hannover, Salzgitter and Gorleben

Interviewed in Hannover in May, 2003: Dr. Detlef Appel, geologist, member of AkEnd.
　　Interviewed in Salzgitter: Dr Bruno Thomauske, BfS and member of AkEnd.
　　Interviewed in Gorleben: Peter Ward, DBE; H. Schröder, Mayor of Gartow Samtgemeinde; H. Legner, member of Gartow council; Harald Müller, *Kurve Wustrow*, Centre for Education and Networking in Non-Violent Action; Prof. Dr-Ing. Klaus Kühn, TU-Clausthal, member of AkEnd; Enrique Biurrun, DBE Communications.
　　The visit included a tour of the Gorleben mine.

2003 September/October, visit to Kassel, Darmstadt, Gorleben, Hamburg and Hannover

Interviewed in Kassel in September, 2003: Prof. Dr Detlef Ipsen, sociologist, member of AkEnd.

Interviewed in Darmstadt: Michael Sailer, Öko-Institut.

Interviewed in Gorleben: Peter Ward, DBE Information; Jürgen Auer, GNS Communications; Marianne Fritzen, campaigner; Wolfgang Ehmke, campaigner; Asta von Oppen, campaigner; Lilo Wollny, former Green MEP; Eckhart Kruse, Pastor at Gartow; Thomas Hauswaldt, lawyer and campaigner.

Interviewed in Hamburg: Nikolaus Piontek, lawyer; Susan Ochse, Greenpeace.

Interviewed in Hannover: Rebecca Harms, Green MEP, later Leader of Greens European Parliament.

2014 April, visit to Bonn, Darmstadt, Braunschweig, Salzgitter, Wendland and Hamburg

Interviewed in Bonn, April, 2014: Anika Limbach, Susanne Matthes, Steffen Patzer and Herbert Hoting, campaigners with *AntiAtomBonn;* Georg Arens, senior official, BMUB.

Interviewed in Darmstadt: Michael Sailer, CEO, and Gerhard Schmidt, scientist, at Öko-Institut.

Interviewed in Braunschweig: Jörg Mönig, Head of Final Repository Research, GRS.

Interviewed in Salzgitter: Michael Müller, BfS.

Interviewed in Wendland: Peter Ward, DBE Geo Information and Chairman of Workers' Committee; Bories Raapke, Managing Director, DBE; Rüdiger Kloth, Public Relations, GNS; Hans-Udo Maury, member of Lüchow-Dannenberg Landkreis, Gartow Samtgemeinde and Gartow councils; Klaus Hofstedder, Bürgermeister Gorleben; Friedrich-Wilhelm Schröder, Bürgermeister Gartow Samtgemeinde; David Beecken, member of Landkreis; Wolfgang Ehmke, campaigner; Asta von Oppen, campaigner; Graf Andreas von Bernstorff, landowner; Graf Friedrich von Bernstorff, landowner.

Interviewed in Hamburg: Heinz Smital, Thomas Breuer, Susanne Neubronner, Conny Deppe-Burghardt, Greenpeace.

List of abbreviations

BBU — Bundesverband Bürgerinitiativen Umweltschutz (Federal Association of Citizens' Initiatives on Environmental Protection)

BfS — Bundesamt für Strahlenschutz (Federal Office of Radiation Protection)

BI — Bürgerinitiativen (local citizens' initiative group)

BLG — Brennelementlager Gorleben GMbH (technical installations of fuel element storage at Gorleben operated by GNS)

BMUB Bundesministerium für Umwelt, Naturschutz, Bau und Reaktorsicherheit (Federal Ministry for the Environment, Nature Conservation, Building and Nuclear Safety)

BMWi Bundesministerium für Wirtschaft und Energie (Federal Ministry for Economic Affairs and Energy)

BUND Bund für Umwelt und Naturschutz Deutschland (Union for the Environment and Nature Conservation Germany, Friends of the Earth)

DBE Deutsche Gesellschaft zum Bau und Betrieb von Endlagern für Abfallstoffe mbH (The German Society for the Construction and Operation of Waste Repositories, DBE Technology operator of Gorleben mine)

DWK Deutsche Gesellschaft für die Wiederaufarbeitung von Kernbrennstoffen (German Corporation for Nuclear Reprocessing)

GNS Gesellschaft für Nuklear-Service mbH (operator of radioactive waste storage services)

GRS Gesellschaft für Anlagen- und Reaktorsicherheit mbH (a Third Sector Organisation TSO non-profit making research on nuclear safety)

Notes

1 CASTOR Cask for Storage and Transport of Radioactive Materials.
2 Konrad is initially expected to take 300,000 m^3 of non-heat generating wastes, 95% of the volume and 1% of the radioactivity when it becomes operational during the 2020s.
3 It is interesting to compare Germany with the other three countries covered in this book. In France, according to the Globescan poll, there was a similar, if slightly lower shift against new build from 66% in 2005 to 83% in 2011. In the USA the percentage favouring new build did not change (40% and 39%) while in the UK there was a marginal increase in favour (33% to 37%). Although this is only one poll and reflects the immediate post-Fukushima situation, the point to note is the extremely strong and continuing antipathy to nuclear energy in Germany.

References

AkEnd (Arbeitskreis Auswahlverfahren Endlagerstandorte) (2002a) *Selection Procedure for Repository Sites, Recommendations of the AkEnd Committee on a Site Selection Procedure for Repository Sites,* BfS, Salzgitter, Germany.

AkEnd (2002b) *Site Selection Procedure for Repository Sites, Recommendations of the AkEnd Committee on a Site Selection Procedure for Repository Sites,* BfS, Salzgitter, Germany.

Aparacio, L. (ed.) (2010) *Making Nuclear Waste Governable, Deep Underground Disposal and the Challenge of Reversibility,* Paris, Springer and Andra.

Appel, D. (2006) 'Historical background of decision making for repository projects in Germany: example for missing participation of stakeholders', in NEA, *Disposing of Radioactive Waste: Forming a New Approach in Germany* FASC Workshop Proceedings Hitzacker and Hamburg, Germany, 5–8 October, 2004, Paris, NEA.

BBC (2011) 'Nuclear power gets little support world wide', *BBC News Science and Environment,* 25 November.

Beck, U. (1992) *Risk Society: Towards a New Modernity,* London, Sage Publications.

BfS (2014) *Gorleben Central Interim Storage Facility*, 24 February.

Blowers, A. and Lowry, D. (1997) 'Nuclear conflict in Germany: the wider context', *Environmental Politics*, 6, 3, Autumn, 148–155.

Blowers, A., Dutton, M. and Warren, L. (2008) *The Overseas Experience of Radioactive Waste Management*, Committee on Radioactive Waste Management (CoRWM), Document No. 2213.1, June.

BMU/BfS (2002) *National Report on Main Nuclear Developments and Regulatory Issues in Germany*, Presented to Meeting of Nuclear Regulators Working Groups, Brussels, 25 November.

BMWi Federal Ministry of Economics and Technology (2008) *Final Disposal of High-level Radioactive Waste in Germany – The Gorleben Repository Project*, BMWi, Berlin, October.

Brand, K.-W. (1999) 'Dialectics of institutionalisation: the transformation of the environmental movement in Germany', *Environmental Politics*, 81, 35–58.

Christiansen, E. (1998) *The Northern Crusades*, Harmondsworth, Penguin, 2nd Edition (first published, 1980).

CoRWM (Committee on Radioactive Waste Management) (2006) *Managing Our Radioactive Waste Safely: CoRWM's Recommendations to Government*, London, CoRWM, November.

CoRWM (2007) *Moving Forward: CoRWM's Proposals for Implementation*, London, CoRWM document 1703, February.

Damveld, H. and Bannink, D. (2012) *Management of Spent Fuel and Radioactive Waste: State of Affairs, A Worldwide Overview*, Nuclear Monitor, 746/7/8, Amsterdam WISE/NIRS, May.

Doherty, B. (2002) *Ideas and Action in the Green Movement*, London, Routledge.

Dominick, R.H. (1992) *The Environmental Movement in Germany: Prophets and Pioneers, 1871–1971*, Bloomington, Indiana University Press.

Dryzek, J., Downes, D., Hunold, C., Schlosberg, D. and Hernes, H-K. (2003) *Green States and Social Movements: Environmentalism in the United States, United Kingdom, Germany and Norway*, Oxford, New York, Oxford University Press.

Ethics Commission (2011) *Germany's Energy Turnaround – A Collective Effort for the Future*, presented by the Ethics Commission on a Safe Energy Supply, Berlin, 30 May.

HMSO (1990) *This Common Inheritance: Britain's Environmental Strategy*, Cm 1200, London, HMSO.

HMSO (1995) *Review of Radioactive Waste Management Policy*, Cm 2919, London.

Hinchliffe, S. and Belshaw, C. (2003) 'Who cares? Values, power and action in environmental contests', in Hinchliffe, S., Blowers, A. and Freeland, J. (eds), *Understanding Environmental Issues*, Chichester, Wiley and Milton Keynes, The Open University.

Hinchliffe, S. and Blowers, A. (2003) 'Environmental responses: radioactive risks and uncertainty', in Blowers, A. and Hinchliffe, S. (eds), *Environmental Responses*, Chichester, Wiley and Milton Keynes, The Open University.

Hocke, P. and Renn, O. (2009) 'Concerned public and the paralysis of decision-making: nuclear waste management policy in Germany', *Journal of Risk Research*, 12, 7–8, October–December, 921–940.

Hugues, P. (1998) *Le Bonheur Allemand*, Paris, Seuil.

International Panel on Fissile Materials, IPFM (2011) *Managing Spent Fuel from Nuclear Power Reactors: Experience from Around the World*, IPFM, September.

Ipsen, D. (2003) 'Public participation and regional development at a nuclear waste disposal site', University of Kassel, Dept. of Architecture, Urban and Landscape Planning, unpublished paper.

Kommentus (2001) *Responsibility, Equity and Credibility: Ethical Dilemmas Relating to Nuclear Waste*, Stockholm, Kommentus Förlag.

Koopmans, R. (1995) *Democracy from Below: New Social Movements and the Political System in West Germany*, Boulder, Westview Press.

Mertens, A. and Haas, A. (2005) *Regional Unemployment and Job Switches in Germany – An Analysis at District Level*, New Economics Papers, European Regional Science Association, Vienna.

Renn, O. (2008) *Risk Governance: Coping with Uncertainty in a Complex World*, London, Earthscan.

Röhlig, K-J. (2013) *The German Programme for the Disposal of Radioactive Waste*, Institut für Endlagerforschung, Workshop on Radioactive Waste Disposal, Stockholm, 30 May.

Rucht, D. and Roose, J. (1999) 'The German environmental movement at a crossroads?', *Environmental Politics*, 8, 1, Spring, 59–80.

Rüdig, W. (2000) 'Phasing out nuclear energy in Germany', *Environmental Politics*, 9, 3, 43–80.

Tiggermann, A. (2010) *Gorleben als Entsorgungs- und Endlagerstandort*, study commissioned by Lower Saxony Ministry for Environment and Climate Protection, May.

SRU (German Advisory Council on the Environment) (2011) *Pathways Towards a 100% Renewable Electricity System, Special Report*, Berlin: SRU, October.

Völzke, H. (2014) *Present Status of Spent Fuel Management in Germany, Federal Institute for Materials Research and Testing*, paper presented at INMM Spent Fuel Management Seminar XXIX, Arlington, January.

World Nuclear Association (WNA) (2014) Nuclear Power in Germany, www.world-nuclear.org (accessed 16 April 2016).

6

CONCLUSION

We don't inherit this land from our ancestors, we borrow it from our children'.
(Attributed to Native American Chief Seattle, 1780–1866)

The legacy of nuclear power – a moral issue

The communities at the heart of this study testify to the fact that nuclear power has left not just a physical imprint in the wastes it leaves behind but has a social expression, too, in the communities whose continuing role is to act as guardians of the legacy of nuclear power. And, just as the physical legacy of nuclear energy is enduring in both space and over time, so the social legacy, once established in nuclear communities, continues so long as the wastes are actively managed. Nuclear communities must endure both the physical threat of living with environmental risk and the social stigma that is often associated with proximity to the inevitable end products: the radioactive wastes arising from nuclear activities. Of course, nuclear communities do not and cannot live in a perpetual state of anxiety and dread, nor should they since, as custodians of these wastes, they have an obvious interest in their safe and secure management. Nevertheless, the possibility, however small, of a serious accident or even a catastrophic event that can and probably will occur somewhere, sometime, places these communities at a disadvantage compared to those that do not face the potentially devastating consequences of living with nuclear risk. It is true that there are communities living near hazardous facilities, those vulnerable to extreme natural hazards, seismic and climatic, as well as those engulfed by war that may also experience traumatic and annihilating events. But it is the potentially global scale of nuclear risk and its persistence over time that marks it off from other risks. And, by its association with weapons of mass destruction, nuclear risk, uniquely, suggests the possibility of global annihilation. For, in its extreme sense, a nuclear accident somewhere, is a nuclear accident everywhere.

Certainly, Chernobyl left a global imprint and Fukushima brought about evacuation of a wide area and the shut down of reactors in Japan and many other countries, some of them permanently.

Nuclear's uniqueness as a risk, a subject of study and an object of policy stems from the fact that it is respected and regarded as somehow different and distinctive from other areas of risk. Radioactivity is pervasive yet invisible, with impacts on health that are known but also difficult to identify and calculate because of the timescales, pathways and intervening variables that result in assumptions and calculations of probabilities so wide in range as to become virtually meaningless. It is not the fact that it is, but that it is thought to be, unique that characterises nuclear risk. For this reason nuclear activities have achieved a privileged status politically in terms of regulation, legislation and policy making. Likewise, nuclear communities, especially those responsible for managing the nuclear legacy that are the subject of this book, have received recognition for their unique social role in looking after the nuclear legacy. They have a moral claim on society for bearing a burden that others are happy not to carry. This moral claim is expressed in terms of equity both spatial and temporal. There is the claim of intra-generational equity, that is fairness between places; and there is intergenerational equity, fairness between generations. These moral dimensions are implicit (and sometimes explicit) in the development of policy making for the management of radioactive waste as we shall see.

The geography of the legacy of nuclear power

By and large the geography of the legacy of nuclear power is relatively fixed. In the early days sites were chosen in remote and economically marginal areas. The nuclear legacy is scattered around the world, in uranium mining tailings, in production facilities such as the centres for reprocessing or at the sites of operating or decommissioned power stations, weapons manufacturing establishments, research facilities and radioactive waste stores. Once established, the geography of nuclear's legacy is tenacious, firmly fixed to those places mainly developed in the early years of nuclear expansion. Later efforts to create facilities for radioactive waste management in greenfield locations encountered resistance and usually failed. In the United States, the site at Yucca Mountain, Nevada, identified as the potential site for the nation's first repository to take the spent fuel stored in sites across the country, has met with successive legal, regulatory and constitutional challenges becoming a political yo-yo, doomed, it seems, to perpetual suspended animation. After years of procrastination, in 1999 the Waste Isolation Pilot Plant (WIPP), a geological repository on a greenfield site deep in the salt formations at Carlsbad, New Mexico, was finally opened to receive transuranic wastes from the defence industry, two decades after construction began, only to encounter problems of seepage that forced its suspension in February 2014. But, Carlsbad was in a remote area of declining potash industry and so did not meet with much resistance, unlike Gorleben, also a salt formation, where tenacious opposition to development eventually prevailed. In France, Bure, too, is a greenfield location where an

TABLE 6.1 Progress with long term management of spent fuel and high level wastes in selected countries

Country	Policy	Progress
Europe		
Finland	Spent fuel direct disposal	Repository development at Olkiluoto
Sweden	Spent fuel direct disposal	Repository site selected at Östhammar
France	Reprocessing/storage/disposal	Cigéo underground disposal project at Bure
United Kingdom	Reprocessing/storage/direct disposal	Site selection process reviewed after lack of progress in West Cumbria
Germany	Spent fuel/reprocessing overseas waste returns/storage/direct disposal	Site selection process under consideration
Spain	Spent fuel storage	Central store site selected (Cuenca)/research on disposal
Belgium	Reprocessing (ceased)/spent fuel/storage/disposal	Research into clay formations for disposal
Netherlands	Long term storage	Central store
Switzerland	Spent fuel storage/disposal	Central store/underground rock laboratories/research into clay formations for disposal
Czech Republic	Spent fuel/storage/disposal	Geological investigation at candidate sites
Slovakia	Stored on site	Site selection process
Hungary	Stored on site	Research on disposal
Bulgaria	Stored at site/sent to Russia	Preliminary consideration of repository
Romania	Stored on site	Preliminary investigations for repository
Russia	Reprocessing/waste returns from overseas/storage	Reprocessing (Ozersk)/central storage at sites and central store/deep well injection/planned underground research laboratory
Ukraine	Storage at sites/sent to Russia	Preliminary investigations for repository
Asia		
Japan	Storage at sites/reprocessing (delayed)	Reprocessing at Rokkasho (delayed)/site selection process for repository
South Korea	Storage at sites/central store	Site selection process under consideration
China	Storage at reactor sites/central store/proposed reprocessing	Central store and repository development (Gansu province)

TABLE 6.1   continued

Country	Policy	Progress
India	Storage at sites/reprocessing	Research on disposal
Taiwan	Storage at sites	Disposal under consideration
North America		
Canada	Stored at site/direct disposal	Repository siting process in progress
United States	Spent fuel storage at sites/clean-up at defence sites/disposal	Site selection process following failure to proceed at Yucca Mountain/defence wastes disposed of at WIPP (Carlsbad)

Notes
1   Table indicates basic policy for management of spent fuel and high level wastes (HLW) from reprocessing in 23 countries in 2016.
2   Most nuclear countries have opted for deep geological disposal as the long-term method but progress varies.
3   Disposal facilities for lower level wastes are in operation or planned in all countries.

underground research laboratory has been developed and which is the presumptive (but not yet conclusive) site for a deep disposal facility. Elsewhere, progress in finding sites for deep repositories has been slow. In the UK, a process of site selection based on voluntarism and partnership stalled when it was decided not to proceed at Sellafield, where the bulk of the country's nuclear wastes are already stored. The process, still based on voluntarism, has to begin again. Similarly, in Germany, the opposition to Gorleben has resulted in a new process of site selection being started. Elsewhere, in Europe, in the Far East (Japan, South Korea, Taiwan) the urge to find sites for deep repositories has resulted in a range of processes that have encountered varying degrees of protest and participation.

Where progress has been made, as in Finland and Sweden, it has been through a process of comparative site evaluation eliminating all but those sites in existing nuclear locations. In Finland the choice has fallen on Eurajoki, where there are two reactors at Olkiluoto and a third under construction. In Sweden, the chosen site is at Östhammar, near the Forsmark nuclear power station and a purpose built repository for intermediate level wastes already constructed under the Baltic Sea. Östhammar was ultimately given preference on grounds of slightly better geological conditions over Oskarshamn, another nuclear complex, where reactors, a central store for spent fuel, an underground laboratory and a future encapsulation plant are located. The tendency in almost every country is to proceed slowly and to focus on those locations where nuclear facilities are well established. In those countries where greenfield sites have been chosen, they have emerged as the preferred sites only after other candidates have been eliminated as unacceptable either for political or scientific reasons or both. Thus, Yucca Mountain and Bure are, in effect, the lowest common denominators in their respective countries, where political opposition has not, so far, been overwhelming and where geological conditions are potentially favourable.

And, even these sites are by no means certain to become the national repository and, like all the other potential sites, face many years of scientific investigation and political procrastination before they are ever likely to open to receive wastes.

Despite all the effort and the apparent priority that the nuclear industry and governments are investing in the search for a permanent solution to the problem of managing nuclear wastes in the future, the real problem is here and now. These dangerous materials are stored in varying conditions of safety and security and are unlikely to be moved elsewhere or put into a repository for decades to come. Even if sites for repositories are found and developed they will not be available to take wastes for decades. Indeed, it is highly likely, inevitable even, that highly active wastes will remain in scattered locations until well into the next century especially if new generations of nuclear power stations are built. The idea of legacy wastes, let alone wastes from new build, being neatly and routinely packaged and transferred to a welcoming and pristine repository there to be entombed for ever, is, with rare exceptions, little more than a fantasy at this point in time. Sites have to be found, safety cases made and complex logistical operations developed before any dreams of a final solution can be fulfilled. And that will not happen any time soon. The reality is that the long-term solution, at least for the foreseeable future, is storage. And, as I shall show, that is where the emphasis of policy making for the foreseeable future should be placed.

Most of the sites where nuclear wastes are stored have not arisen as the outcome of any strategic siting strategy. Rather they are the outcome of specific decisions taken on the basis of a few basic geographical criteria. Most of the radioactive waste sites were the consequence of siting decisions for nuclear power stations, reprocessing works or military reservations. They are where they are, not because they were deliberately selected as the most suitable places to store or dispose of wastes but because wastes are an inevitable adjunct to the production of energy, nuclear fuels or nuclear weapons. Although they were not purposely selected, they will remain as the locations for the nuclear legacy long after the original purpose of the sites has vanished. Even in the few cases where sites for the storage of wastes or for repositories in which to bury them are the outcome of siting strategies, the locations have usually been determined on geographical and scientific (geological) criteria before any public and political consultation has been undertaken. In those instances where a voluntary process of site selection has been introduced, it has either been restricted to making a choice between a few sites, as in Sweden or Finland, or has relied entirely on the idea of a 'white map' as in the UK or Germany which has, so far, got nowhere beyond not (yet) landing Sellafield or Gorleben. The history of trying to find sites for a repository or central location for radioactive wastes is littered with examples where, to transcribe a biblical expression, many sites have been called but few chosen.

The aim of this book has been to try to explore the relationships between the nuclear legacy and local communities. In this final part I want to look back to identify some common geographical, political and social themes which have emerged from a study of four nuclear communities. But, the experience of these communities also enables me to look forward to reach some conclusions about how society should approach the problem of looking after the legacy of nuclear power in the future. Three overarching conceptual ideas can be drawn out of my analysis and form the basis for my concluding reflections. These are: first, peripherality and inequality, exploring the nature of these communities; two, peripheralisation and power relations, explaining the political geography of the nuclear legacy; and, three, responsibility and equity, raising principles of how we should manage the nuclear legacy in the future.

Periphery and inequality

The first conceptual idea concerns the relationships that constitute those communities living with the legacy of nuclear power. This is an issue that focuses on the nature of peripheral communities and the inequalities that define them. The focus of attention of this book has been on five locations in four countries which are associated with the long-term management of the most highly active and potentially dangerous materials of nuclear's legacy. While each has its own narrative explored in separate chapters, all five express in varying ways what I have termed 'peripheral' characteristics which have been created and reinforced by processes of 'peripheralisation', the processes of push and pull whereby the nuclear legacy is confined to established locations and is generally repelled from colonising new ones. Although the five communities are, perhaps, archetypes of peripheralisation, they do suggest some features that explain more broadly why the geography of the legacy of nuclear power has, by and large, become ossified and seems likely to remain so.

Geography – places on the periphery

Hanford and Sellafield had their origins as plutonium factories making the material for the nuclear arsenals that were piled up in the arms race during the Cold War. They were shrouded in secrecy, places where danger lurked, deliberately chosen for their remoteness. While Hanford remained predominantly a military reservation, Sellafield became a complex where military and civil nuclear operations coexist. In common with La Hague, which came later and which became the focus of reprocessing for the French nuclear energy programme, all three sites are on the territorial edge of their countries, in remote areas where farming and fishing were the traditional occupations. Gorleben and Bure have a different provenance. Bure is a site chosen almost by default, by a process of elimination as other potentially suitable candidate sites fell by the wayside, removed as a result of unsatisfactory geology or opposition. A foothold has been gained and various procedural and regulatory milestones have been passed and so Bure may lead on, in time, to become France's first national repository. By contrast, there seemed to be a combination of scientific, social and economic criteria used to identify the Gorleben site but the lack of consultation or comparative assessment rendered the selection obscure and contributed to the sense of injustice that was such a motivating force to the Gorleben movement. Like Bure, Gorleben has an underground laboratory but its future as a repository seems, now at least, far more doubtful. Both sites are in what may be called the 'internal periphery' rather than on the edge of the country, though Gorleben was originally on the border with East Germany and Bure is not so far from the German border. But all five locations are peripheral in terms of remoteness and distance from the mainstream. It is this relative geographical isolation that provides some protection and insulation from external processes.

Economy – places at the economic margin

A prominent factor in the original location of the nuclear legacy is economic marginality, places experiencing underdevelopment or economic decline. Once established the common economic denominator of these communities is their monoculturalism, that is, a dependence on a dominant activity, in this case managing the nuclear legacy. As at Hanford and Sellafield and increasingly, too, at La Hague, this may involve large nuclear complexes with a workforce of more than 10,000 employed in a range of activities including reprocessing, spent fuel management, disposal (of lower activity wastes), decommissioning and clean-up. At all three sites the workforce has been stable in recent years though there is the prospect of a slow decline in the longer term. All three receive substantial and continuing state support. At Sellafield and Hanford this is to ensure the safe and secure management especially of the tanks and ponds which present serious risks and difficult technical challenges. At La Hague the original rationale for reprocessing as an integral part of the closed fuel cycle may have changed but processing spent fuel will continue into the future primarily as a method of managing waste through storage and vitrification. The other two sites, Bure and Gorleben, are still largely greenfield though both have underground laboratories and the latter has a spent fuel store and a mothballed encapsulation facility. The nuclear workforce is small, declining at Gorleben and increasing at Bure. In these places the nuclear component in the local economy is small and its impact relatively weak though potentially growing as at Bure.

At the three major sites (Hanford, Sellafield and La Hague) dependence on the nuclear industry makes them at once both vulnerable and resilient. The guaranteed state investment in legacy management provides a reasonably secure future, at least for those working in the industry, in its local supply chain and in the services supported by the income which flows into the local economy. At the same time the dominance of the nuclear industry with its relatively high pay rates creates inequalities and may result in a lack of inward investment in other activities. This may retard diversification as is the case around Sellafield. On the other hand, the nuclear industry may stimulate other related activities, as at Hanford, and become an anchor within a far more diverse economic landscape as in the Tri-Cities. In most nuclear communities dependence has not meant exploitation, impoverishment or vulnerability as is often the case in monocultural communities. Instead, though economic and social inequality may be evident, these nuclear communities experience a sustained economic base supported by the state. In the sense that they enjoy comparative economic well-being and security they well deserve the epithet 'Plutopia' accorded to them in the book of the same name (Brown, 2013).

Cultural resignation and resilience

In a previous chapter (Chapter 3) I analysed at some length the somewhat ambiguous and sometimes contradictory values and attitudes manifested in West Cumbrian relationships to the nuclear industry. The social characteristics described in Sellafield, though culturally specific and distinctive, are in broad terms an integral component of peripherality and are reproduced in general, though locally specific, form in the other developed nuclear communities I have studied. It is difficult to summarise the main features of such complex communities and I would refer to the insights provided in the studies of nuclear societies such as Loeb's *Nuclear Culture* which deals with Hanford, Wynne et al.'s sociological study of perceptions in West Cumbria or the oral histories recorded for Hanford, La Hague and Sellafield which relay a wealth of individual insights and perspectives (Loeb, 1986; Wynne et al., 1993; Zonabend, 1993; Sanger, 1995; Davies, 2012). From these sources and from interviews in all the communities studied in this book I would pick out three distinguishing and complementary cultural features of these nuclear communities. First is a *realism*, a recognition of their role which translates into a sense of pride and patriotism (certainly in the case of Hanford) that is subdued rather than arrogant. It gives rise to the feeling that these communities are special and that the obligations they fulfil for society should be reciprocated in appropriate compensation. Second is a *resignation*, recognition that the work they undertake is dangerous, unglamorous and stigmatised leading to fatalistic self-deprecation. This may reinforce a negative image and encourage claims that these places make for improvement, diversification and community development. Third, there is a *pragmatism*, a perception that these communities are resilient and able to adapt and adjust to changing circumstances. They are at once both flexible and obstinate, resisting change while also ultimately embracing it, providing such communities with a fortitude and unity of purpose that contributes to their continuing survival. These same features but in different forms may also explain why in some places the nuclear industry has failed to gain a real foothold. Gorleben provides the evidence. There the community was realistic in the sense of perceiving itself to be culturally specific (the idea of the Wendland) but, in this case, was not fatalistically resigned to hosting an industry which would denigrate the positive cultural image of Wendland. Certainly, at Gorleben the community showed a high degree of pragmatism and resilience in its successful campaign to protect its cultural features.

Political power and powerlessness

It would be rational to conclude that the geographical remoteness, cultural realism and resignation and economic marginality of nuclear oases would be compounded by political weakness. And, in several respects, this appears to be so. Decision making for the nuclear industry tends to be centralised in state run or supported institutions. Strategic decisions of fundamental importance to the local nuclear industry and communities are taken at the level of state corporations or government departments or even international organisations. Thus, the overall budget for Sellafield's operations is set by the UK government and administered through the Nuclear Decommissioning Authority (NDA), a nationalised body responsible for managing the civil nuclear legacy country-wide. Similarly, the annual subventions for Hanford's clean-up are settled in Washington DC. La Hague has always been a component of predominantly state-owned nuclear companies (see Chapter 4) and the Bure facility is the responsibility of Andra, the French waste management authority. In Germany, too, federal government owned enterprises DBE and BLG manage the underground salt mine and the waste storage facility. There is, of course, some devolution of decision making though this may create fragmentation rather than a unified voice at local level. Contractorisation, especially at Hanford and to a lesser extent at Sellafield where the NDA maintains a strategic coordinating role, may, in some cases, lead to differences over the methods and timescales of clean-up.

On the other hand, the significance of these legacy industries in their local context combines with their peripheral conditions to enhance their political power. In the case of Sellafield and Hanford the recognition that there is an obligation to these communities as well as a pressing environmental requirement to deal with contamination of the ponds and tanks as well as the rest of the primordial mess has bequeathed a moral imperative on the nation. The practical outcome has been a virtually guaranteed budget representing the nation's gratitude to those places that bear the burden of the nuclear legacy. The amount of expenditure on cleaning up Sellafield, though huge (annual cost £1.7 billion and total clean-up cost estimated at £53 billion in 2015) is relatively uncontested for it often seems to its critics a feature of the nuclear industry writ large that what it asks for, it gets. In a sense the nuclear legacy industry is the acme of non-decision making; it is powerful by reputation, proactive in its claims but passive in achieving successful outcomes (Crenson, 1971). In France, too, there has for long been relatively unquestioned support at the political level for reprocessing which has been accepted as an integral and essential component of the French nuclear cycle. Nevertheless there has been a suspicion that EDF, the main customer, did not want to have to use the plutonium from reprocessing and, consequently, La Hague focused on dealing with its foreign customers first. Although the role of reprocessing in terms of necessity, scale and cost has been increasingly questioned, it has not, yet, reached the point of policy reversal. Even though France is committed to scaling down its nuclear industry, it is unlikely that it will give up its commitment to reprocessing, rather it will survive as a key component of nuclear waste management. At Bure the industry is not yet sufficiently large nor embedded locally to exercise significant power and decisions remain at the centre. In the Wendland, the nuclear industry has never gained sufficient local scale or employment to expand beyond its bridgehead operations.

In all four countries nuclear interests have been able to captivate if not entirely capture the local representative institutions. The nuclear industry continues to exercise political influence in the Tri-Cities and over the consultative bodies such as the Hanford Advisory Board. In West Cumbria, Copeland Borough Council has long been in thrall to Sellafield, neighbouring Allerdale Borough less so. Similarly, La Hague has strong support in the nuclear peninsula and Bure has gained a foothold in some of the small communes as well as regional support in the two departments (Haute-Marne and Meuse) which it straddles. The communes around Gorleben have also been supportive of the industry. But, as distance from these nuclear complexes increases, so support begins to fade away. The political geography reflects a 'doughnut' effect with a nuclear oasis strongly pro-nuclear surrounded by areas where the population is generally indifferent to nuclear issues and, occasionally, anti-nuclear sentiments may flourish. Thus, pro-nuclear Tri-Cities are within Washington, a state where concerns about Hanford are expressed, if at all, in cities many miles from Hanford. West Cumbria seems to be a quintessential nuclear haven but beyond, in the rest of Cumbria, there is political ambivalence towards Sellafield, an economic asset but environmental hazard. While La Hague enjoys a passive acceptance in the northern Cotentin, further south the preoccupations of a rural agricultural economy predominate. The impression of the nuclear industry on the politics of the region around Bure is in the nature of an interloper, relatively immature, developing and fluid in the process of converting a greenfield site into a nuclear oasis.

Gorleben stands out as the exception, an area where from the very outset the nuclear industry was made unwelcome, opposed and eventually halted. Gorleben is the nuclear community that never was. The reprocessing and waste management complex at Gorleben was promoted decades after Hanford, Sellafield and La Hague had become established, at a time of controversy over nuclear energy emphasised by the TMI disaster and especially acute in Germany where the association with nuclear weapons at the height of the Cold War provoked public concern and a powerful peace movement. Moreover, Gorleben appeared, as it were, out of the blue, imposed on an unsuspecting but also nuclear sensitive population. The remarkable coalition of anti-nuclear forces that mobilised, revivifying a long dormant idea of a cultural 'Wendland' as a focus for defending their territory against the nuclear imposter, proved imaginative, flexible, tenacious and single-minded.

The longer the nuclear industry has been established the more influence it has within its community. Hanford and Sellafield have developed over a long period and become settled and accepted among the population. This is true, too, of La Hague, established later and still an operating site. In all three places community and industry have a mutual interest in defending jobs and in applying leverage, actively or passively, on government to secure support for nursing the nation's legacy of wastes. In complete contrast is Gorleben, a place where the nuclear industry failed to get fully established and where its leverage was feeble, lacking a strong local base and consistent political commitment from federal and regional government. Confronted by an increasingly powerful coalition of anti-nuclear forces, the Gorleben project never got fully off the ground. These different histories reinforce one of the key conclusions of this book that, by and large, the geography of nuclear's legacy is largely established. The places which guard the legacy of the past are those most likely to continue doing so. It may be that some greenfield sites can be established, WIPP and Bure being possible examples. But, as Gorleben has shown, it may be extraordinarily difficult if not democratically impossible to find new locations today which will be willing hosts for managing the nuclear legacy and may well resist having it imposed upon them. And, as Sellafield has shown, it may prove difficult even to establish deep repositories for managing the nuclear legacy in the nuclear industry's heartland.

Peripheralisation and power relations

I come now to the second conceptual idea concerning power relationships. As I have just explained, nuclear communities that are hosting the nuclear legacy manifest inequalities both internally and externally within and between communities. Most of all these places experience inequalities of risk, places which have a disproportionate share of the risk to environment, health and well-being that is associated with proximity to large volumes of radioactive materials. Although nuclear communities are unequal as a consequence of their physical location in what may be termed 'landscapes of risk' and may suffer the social stigma of living with risk, they are neither entirely powerless nor necessarily economically disadvantaged (Blowers, 1999). Indeed, they often possess an economic stability and power of political leverage that is routinely denied to communities living with hazardous activities in often declining industrial sectors.

Nuclear communities are a product of power relations changing over time in response to changing discourses and the mobilisation of resources. They have been created and developed by dynamic processes of what I have earlier called 'peripheralisation', the motions of push and pull which have both attracted and repelled nuclear activities to and from the periphery. In the Introduction (Chapter 1) I described how this process of peripheralisation works and applied it in general terms to nuclear communities. Having now examined four (or five if Bure is included) nuclear communities in depth and in detail I have the empirical evidence to confirm the thesis of peripheralisation, to show how peripheralisation is an expression of power relations creating and reinforcing the physical and social components that constitute the legacy of nuclear power. Throughout this book I have nowhere attempted a precise definition of 'community' although I am aware from earlier work I undertook that a myriad of attempts have been made, including such seminal works as *Gemeinschaft und Gesellschaft* by Ferdinand Tönnies (1887) and Max Weber's *Economy and Society* (1921). The problem is that attainment of a sociologically satisfying definition of the idea of a nuclear community incorporating the various dimensions of 'community' together with an appropriate spatial expression would be comparable to a successful search for the Holy Grail. And it is not necessary for the purposes of this study to attempt such an elusive goal. However, I have focused on the geographical, economic, social and political dimensions that, in broad terms, characterise what I have called 'peripheral' communities.

It is these characteristics that help to explain how peripheralisation works and animates the power relations that create (or prevent) the places that host our nuclear legacy. Earlier I introduced the idea of power as discourse and indicated how changing nuclear discourses over time influenced the balance of power relations and the ability of communities broadly defined to deploy resources of power to advance or defend their interests. In the intervening chapters I tried to show how discourses have tended to shape and change the individual fortunes of particular nuclear communities. As discourses have changed over time so, too, have the resources of power available to nuclear communities. In some cases there has been a predominant discourse, in others competing or overlapping discourses and in others a succession of discourses. Consequently, some nuclear communities have enjoyed relative consistency in the balance of power relations providing stability while others have experienced more fluctuation expressed in conflict over development. In all the cases studied prevailing discourses shifting over time have left their imprint on the nature of the legacy and the communities which manage it.

The early nuclear history was dominated by what I called a discourse of Trust in Technology. This discourse attained its apogee in the years immediately after the Second World War when nuclear science was deployed mainly in the development of nuclear weapons. Thus, Hanford and Sellafield and other sites, notably in the USA (Oak Ridge, Tennessee and later Savannah River Site, South Carolina) and in the Soviet Union (in the Urals in the Chelyabinsk region and near Krasnoyarsk in Siberia for example) were secretive and inaccessible locations, islands of high security defended by barriers, fencing and armed guards. They were virtually unknown to the outside world or even, for that matter, to those inside. At Hanford, for example, despite having 50,000 workers on site during the war years, no one, apart from a small scientific and political elite, had any idea what was going on. Even as awareness developed, these nuclear activities nurtured an aura of scientific mystery and inculcated a sense of awe and wonder on the part of an unquestioning public. Throughout these early decades of nuclear development there was a tendency for deferential trust in expertise and high expectations from the new technology (Falk, 1982; Pringle and Spigelman, 1982; Camilleri, 1984). Nowhere was this more in evidence than in France where the nuclear industry entered the scene during the post-war era of recovery of national pride placing faith and commitment in French technology. Nuclear energy became the very epitome of this surge in pride, 'le rayonnement de la France' capturing the sense of aspiration to technological leadership (Hecht, 1998). It is especially in France that the discourse of Trust in Technology, though challenged and diminished, has survived, imparting a sense of stability and security to La Hague and the local community. At Hanford the discourse has shifted from one of pride in achievement to commitment to clean-up conferring stability and continuity on the community that supports and is, though to a lesser extent than formerly, supported by it. Sellafield's history has been politically more turbulent but has never forfeited the underlying mutuality of its relationship with the West Cumbrian community.

Sellafield was more exposed than either Hanford or La Hague to the marked reversal of discourse that occurred in the 1970s with the advent of the discourse of Danger and Distrust. This was a period whose beginning and end was marked by major accidents at TMI and at Chernobyl, a period, too, when nuclear programmes were being promoted as a counter to fears of shortages of oil. The nuclear industry found itself increasingly on the defensive. Sellafield was plunged into conflict, first over the THORP reprocessing plant, then over its radioactive discharges and later over the Nirex proposal to develop an underground laboratory as a prelude to the deep disposal of the nation's legacy of radioactive wastes. Earlier revelations about the Windscale accident in 1957 had contributed to undermining the reputation of the plant. For a period Sellafield symbolised nuclear's nemesis as anti-nuclear forces, part of the emergence of a more dissenting and discontented mood in civil society, checked the development of new nuclear power stations and eventually of reprocessing in the UK. But, it was in Germany that the discourse had its most powerful impact feeding a vigorous anti-nuclear power movement. Although it developed too late to impede the German nuclear power programme, it had a decisive impact on reprocessing, stopping proposals at Gorleben and at Wackersdorf and forcing Germany to look elsewhere, to France and the UK, to reprocess its spent fuel. At Gorleben only the underground salt mine and the interim waste store were realised of the original *Entsorgungskonzept*; there was no reprocessing and the encapsulation plant though built has never operated. The grandiose Gorleben project never materialised, a victim of the changing power relations as the discourse of Danger and Distrust coursed through Germany. Indeed, in Germany the shift in power relations once established has continued leading on to the ultimate demise of the German nuclear industry. Meanwhile, in the US and in France the anti-nuclear discourse, vigorous in its manifestations of conflict, also had its biggest impact on legacy issues, preventing the selection of various sites for deep geological disposal in France and leading to impasse in the struggle to establish Yucca Mountain as the place for the burial of the legacy wastes in the United States.

As the nuclear industry's advance was halted and had retreated more into its periphery, the conditions were ripe for the emergence of a new, more emollient discourse of Consensus and Cooperation. With the Cold War ended and fears of energy shortages fading, there emerged an opportunity for hitherto opposing forces to find some common ground. The prospects for a less combative relationship between the nuclear industry and its adversaries were especially favourable on the question of managing the nuclear legacy. With expansion apparently over, the nuclear industry and governments could focus more on dealing with its Achilles heel of radioactive waste while anti-nuclear movements would be more than willing to cooperate in solving what they saw as a major threat to health and the environment. The so-called 'participative turn' had ushered in a more open and transparent approach to policy making with an emphasis on deliberative democracy and voluntarism. This was actively applied to relations between the nuclear industry, government and local communities. This potential consensus was especially marked in the UK where, following decades of failing to deal with the problem of nuclear waste, a new initiative, based on a voluntary, participative and partnership approach to finding a solution, was started and embodied in the work of CoRWM (see Chapter 3). This led eventually and inevitably to seeking a site for a repository in Sellafield hard by the stores for the nation's nuclear legacy. In France, too, there was a more conciliatory approach begun with the Castaing Commission to identify potentially suitable locations for a waste disposal facility (see Chapter 4). In the United States there was broad agreement on the necessity for clean-up including the Hanford site and the WIPP facility was opened although Yucca Mountain remained a potential source of future conflict. In Germany the end of the twentieth century had been marked by a political consensus around the phase-out of nuclear energy and the new century began with the AkEnd report, in its way a seminal discussion of a new and participative approach to selection of sites for a repository. The report was elegant but theoretical and was eventually abandoned though its precepts reflect a critical turn in power relationships (see Chapter 5). Overall, the period of Consensus and Cooperation, though it was more marked in some places than others, was a relatively tranquil period in which power relations on the issue of long-term waste management were finely balanced.

Although the discourse of Consensus and Cooperation maintains its influence on processes and policies for managing the nuclear legacy in all countries, it held centre stage, if at all, for a relatively short period. In the first decade of the new century this discourse, though not supervened, became compromised by the emergence, or should I say resurgence, of a more thoroughly pro-nuclear discourse in some countries though not all. Indeed, in the wake of the accident at Fukushima Daiichi in March 2011, some countries opted for phase-out of their nuclear industry including Germany (where the date for phase-out was brought forward to 2022), Switzerland, Belgium, Italy and Taiwan, with France, for example, proposing to reduce its nuclear energy contribution to electricity supply from three-quarters to half the total by 2025. In the United States the accident, together it must be said with the rapid expansion of gas supply from fracking, derailed a stuttering revival of nuclear power. In Japan, Fukushima caused the shut down of all the country's reactors and none were restarted until 2015. By contrast, the UK's ambitious programme of a fleet of new nuclear power stations on eight coastal sites with around 16GW capacity to be deployed by 2025 was scarcely affected by Fukushima. Promise far exceeded performance and, by 2015, it was questionable if even one power station (comprising two reactors with a total capacity of 3.3GW) at Hinkley Point would be delivered around 2025 and there was considerable doubt about the scale of a future programme dependent on foreign investment.

It is difficult to say whether nuclear worldwide is in a state of suspended animation before slow decline or will retain a significant share of the energy mix albeit based on new reactor technologies (Elliott, 2007; Sovacool, 2011). But, nuclear renaissance or not, the stuttering nuclear revival has been supported by a new discourse of Security. This discourse combines four elements: one, national security in the light of 9/11 and subsequent threats of war and terrorism; two, economic security especially following recession; three, energy security supporting the need for a mix of energy sources; and four, environmental security, the development of low carbon energy to meet the challenge of climate change. It is in terms of energy and environmental security that nuclear power is justified though its claim is contested. The issue of dealing with its legacy tends to be discounted yet it is here that the industry is most vulnerable. The future development and security of nuclear energy rests on its ability to find an acceptable and safe permanent solution to the legacy problem. Yet finding somewhere, anywhere, that is both scientifically safe and socially acceptable has proved an intractable problem as the histories in this book have shown. In the case of Hanford the effort to make the environment safe and secure will depend on a successful clean-up, removal of wastes and returning most of the site to natural land use. At Sellafield, two attempts to secure an underground disposal facility have failed and it is improbable that a further attempt based on voluntarism will be successful here or elsewhere in the near future. In France the future of the nuclear legacy has an impressive Gallic logic of ultimately transferring vitrified and encapsulated wastes from La Hague and other sites to a deep disposal facility not finally decided but focused on Bure. But, it will be many decades before this tantalising prospect comes anywhere near fulfilment. As for Germany, with Gorleben stalled, the process of finding a site has begun all over again with Gorleben still in the picture but likely to be vigorously resisted should the case be revisited. Meanwhile wastes will be stored at nuclear sites around the country. The bald fact is that the nuclear legacy is already being managed where it is, in scattered stores all around the globe and there is very little prospect of it going anywhere else anytime soon. In the age of Security the key issue is not finding places to bury the waste and clear the legacy for that will not happen for decades especially if there is new nuclear development creating a future legacy to be managed. The issue is to ensure the safe and secure management of wastes for the foreseeable future in those places where it already exists. The nuclear legacy is a problem not only for us but for many generations to come. As such it is a moral issue.

It is a moral issue

The third conceptual idea on which to reflect relates the histories of the nuclear communities to the future prospects for the management of the nuclear legacy. By virtue of its associations with nuclear energy, nuclear weapons, the dangers of proliferation and terrorism, the nuclear legacy is a moral issue raising a myriad of ethical concerns. Indeed, the very nature of the legacy poses ethical questions that are profoundly difficult to fathom. The nuclear legacy is, at one level, unevenly distributed impacting most on those places where it is managed and contained. But, once released into the environment whether routinely, accidentally or deliberately, radioactivity is unbounded both in time and space. It is the diffuse and everlasting quality of radioactive risk that makes policy making for management so tantalising and contentious. The pathways that radioactivity can take are subject to uncertainties and, ultimately, to indeterminate or unknowable risks and consequences. In the end the choices cannot be arbitrated by scientific understanding and reasoning alone, they must be understood and, perhaps, validated by recourse to principles and values. As Barbara Adam puts it:

> The timescale and complexity of interactions involved place radioactive waste management to a large extent outside the realm of scientific control, prediction and risk assessment. In contexts of uncertainty, indeterminacy and non-knowledge we are outside the world of verifiable expertise and are instead in the realm of morals and values.
>
> *(Adam, 2007, p.15)*

While this may be true, policy making must be based on scientific evidence as well as societal acceptability. But, science can only establish the empirical possibilities that are practical and achievable necessarily hedged around by uncertainties, what is or will be or might be. Where ethics comes in is to articulate 'sets of principles that tell us what counts as acceptable and unacceptable' (Rawles, 2007, p.26), what we should do, what is right or wrong, good or bad. The problem then becomes one of how to translate ethical principles into choices. In the case of the legacy of nuclear power, the ethical issues may be expressed in terms of equity, what is fair and unfair, to open up the debate about where and how to manage radioactive wastes. There are three broad areas of equity to consider in relationship to making policy choices for nuclear's legacy: procedural, intra-generational and inter-generational. How, where and when?

Procedural equity

During the discourse of Consensus and Cooperation there was considerable emphasis on the idea of procedural equity, that is fairness in decision making processes and procedures. Hence the opening up of policy and decision making to wider and inclusive participation characteristic of what social scientists fondly referred to as the 'participative turn' (see Chapter 1). What became especially fashionable was the idea of 'deliberative democracy', an approach to political debate familiar in the classical world but refined and developed in the late twentieth century especially in Germany at a philosophical level by Jürgen Habermas (1981), related to environmental issues by Robert Goodin (1992, 2003) and developed in specific applications by Ortwin Renn (2008). Basically, deliberative democracy insists on equality between participants engaged in argument and reflection. Its key features are, therefore, inclusiveness and unconstrained dialogue (Smith, 2003). A battery of techniques has been developed to support and enhance deliberation and it has especially been applied in areas where societal consensus or at least workable agreement is desired (Dryzek, 1997).

In the case of radioactive waste deliberative approaches have been used, most notably by CoRWM in the UK (2005) (Chapter 3). In a less pure sense participative approaches seeking public acceptance of policies for radioactive waste management were encouraged, though not implemented, by the AkEnd in Germany (Chapter 5) and, to an extent, used in France in the so-called Bataille Law following the failure of the Castaing Commission to achieve acceptable comparative sites for a proposed repository (Chapter 4). Perhaps the most significant outcome of the deliberative turn has been an emphasis, in theory if not practice, on the desirability of openness, transparency, participation and partnership in long-term decision making for radioactive waste.

Participative approaches imply a consensus approach to decision making and have been used to support, though not supplant, the role of representative democracy in taking decisions. Ideally, conclusions arrived at through participation and partnership will produce recommendations that carry public consent and which are then endorsed and legitimated by an elected representative body. This was the model espoused by AkEnd and in the UK it was put into practice through the MRWS process in West Cumbria. But, this process encountered two fundamental flaws. One was that deliberative democratic processes tend to be open-ended and consensual conclusions may be insufficiently robust to be applicable as policies. Or, hard and difficult choices may be avoided in the effort to achieve agreement. Deliberation, almost by definition, does not have a decision rule. 'In a pluralistic world, consensus is unattainable, unnecessary and undesirable. More feasible and attractive are workable agreements in which participants agree on a course of action but for different reasons' (Dryzek, 1997, p.170). The other, consequential, flaw is that room left for interpretation may be seized upon by representative bodies to draw their own conclusions based on the interests they represent.

The UK illustrates this problem perfectly well. At the level of generic policy making, CoRWM prepared an interdependent set of recommendations supporting deep disposal as the preferred approach to radioactive waste management based on a programme of public and stakeholder engagement which was, in broad terms, endorsed by the national government. However, the attempt to apply the model of voluntarism and partnership to finding a possible site for a repository in West Cumbria was fatally inconclusive and, once exposed to the conflicts in representative democracy, the process stalled. From experience so far it may be said that deliberative approaches to decision making for radioactive waste are, at best, helpful for formulating broad policies based on consensus; at worst they may become vehicles for procrastination. These processes flourished during the period of Consensus and Cooperation when the power relations between the nuclear industry, government and civil society were more balanced. They have diminished as concerns with security have arisen. However, given the complexity of the issue, the range of potential options and the immense timescales involved, unconstrained dialogue and indecisiveness may prove to be a more acceptable way forward than premature legitimation of policies for radioactive waste that may foster nuclear expansion but at the expense of alienating communities burdened with the nuclear legacy now and in the future. When policy making shifts from the generic to the specific, from identifying options for waste management to finding sites for facilities, consensus may be difficult to sustain in the face of conflict. Radioactive waste involves not only a conceptual problem of defining the best option for its management but also an empirical problem of identifying a place where it can be managed. And the question of where brings us to the second broad area of equity, fairness in relation to place.

Intra-generational equity

It is axiomatic that the nuclear legacy must be managed somewhere and this book has focused on places, like Hanford, Sellafield and La Hague, which have very significant accumulations of radioactive wastes, as well as Gorleben where plans for an integrated waste management complex have been thwarted. Clearly, the uneven development of legacy facilities represents an inequality in that, in this particular respect, some places are experiencing a burden of risk on behalf of society. They have been distinguished as a type of peripheral community expressing features of economic, social and political inequality. The issue is whether the burden they bear is inequitable; is it fair and, if it is not, should they be compensated? It is difficult, though not impossible, to apply this question retrospectively to those communities already hosting the legacy but it becomes a very pressing question in the search for a site for storing or disposing of radioactive wastes. The intra-generational ethical principle here is that actions should not impose an unfair or undue burden on individuals or groups within the present generation.

In order to provide any guidance for policy making, intra-generational equity may be expressed in terms of criteria for siting radioactive waste facilities. No single criterion satisfies the principle of intra-generational equity; a blending of criteria is necessary to achieve an equitable outcome. Perhaps the best way to solve the conundrum is to recognise that radioactive waste facilities are a burden and that communities hosting them ought to be compensated for doing so. This basic principle of compensation has now been recognised in several of the approaches to siting waste facilities in most countries. It is regarded as a necessary component, combined with partnership and participation to a successful siting strategy. The problem is how to provide compensation as a recognition of responsibility and not as a bribe. Several approaches have been tried with an emphasis on economic regeneration as in France, investment in community benefits as in the UK or the idea of 'perspective compensation', a self-determining regional development concept put forward in Germany (see Chapter 5). From these approaches it seems two principles are emerging which provide both a practical and ethical focus for effective siting strategies. These are:

> Voluntarism – this principle recognises the right of communities voluntarily to opt in to a siting process but, most importantly, it also provides communities the right to withdraw from the process at any point up to a final decision.

Well-being – the principle whereby a community hosting the nuclear legacy is not disadvantaged, rather achieves positive benefits from doing so. The principle was stated in general terms by CoRWM (2007, p.12) as: 'By well being we mean those aspects of living which contribute to the community's sense of identity, development and positive self-image'.

Taken together these principles of voluntarism and well-being offer a broad ethical principle of intra-generational equity whereby local communities exercise greater control over their affairs and develop their own vision for the future. This goes beyond conventional conceptions of compensation in terms of financial investment and community provision towards the notion that any community accommodating part of a country's nuclear legacy should enjoy a perceptible enhancement of its quality of life and thereby become a better place in which to live. This may sound idealistic but it is an aspiration that needs to be operationalised if new sites are to be established. And, in terms of intra-generational equity, it is an aspiration that should be achievable for those communities which already host the nuclear legacy and which will continue to do so. For the principle of intra-generational equity is inextricably linked to intergenerational equity in the sense that the present generation takes decisions that bind future generations and seeks compensation on behalf of those who, as yet, have an interest but no vote or voice. The ethical imperative to enhance well-being applies not just to our present generation but should be a binding commitment to future generations also.

Intergenerational equity

Many of the most active and dangerous radionuclides contained in radioactive spent fuel, high level wastes and other products from the nuclear cycle have very long half-lives, some tending to infinity.[1] Plutonium-239, commonly regarded as among the most dangerous radionuclides, has a half-life of 24,000 years and, of course, remains hazardous during its slow decline thereafter. There are, of course, dangerous radionuclides with very short half-lives (iodine-131, 8 days; strontium-90, 28 years) as well as those extending to millions of years. To get some perspective on this, had it existed it would have taken a radionuclide of plutonium-239 from the last glaciation until the present day to decay to half its life. The point here is that the nuclear legacy poses a risk to environments and human health for periods which extend well beyond our comprehension. This is what generates fear and anxiety and prevails on governments to ensure that the legacy is managed safely to keep present and future generations free from harm. Radioactive waste has long been the Achilles heel of an industry anxious to overcome its negative image and lack of public trust.

The principle of intergenerational equity has been a lodestar of radioactive waste management policy. It is motivated by concern for the future, yes, but also in recognition that cleaning up and making safe the legacy is politically vital to the continuing development of nuclear energy. The principle of intergenerational equity tends to be adumbrated in the language of anthropocentric sustainability. The sustainability principle of the International Atomic Energy Agency (IAEA) is put straightforwardly: 'Radioactive waste shall be managed in such a way that will not impose undue burdens on future generations' (IAEA, 1995, Principle 5).

This is a rather provisional, even equivocal statement but it does mark a recognition of the role of today's guardians of the legacy in protecting society, 'in such a way that predicted impacts on the health of future generations will not be greater than relevant levels of impact that are acceptable today' (Principle 4). There is, then, a presumption of continuity of stewardship for a future presence, a 'duty to preserve this physical world in such a state that the condition for that presence remains intact' (Jonas, 1984). It is this that accounts for the singularity of purpose in the research and development of scientific safety cases designed to demonstrate that geological disposal can safely entomb highly active wastes and ensure barriers, engineered and geological, that will prevent radionuclides reaching the accessible environment for tens of thousands of years. The highly expensive and technologically complex efforts to vitrify and encapsulate wastes that are being undertaken at Hanford, Sellafield and La Hague in preparation for ultimate burial indicate just how seriously the problem of safe management is taken.

Paradoxically, the principles that inspire us to secure the future are being countered by uncertainties that lead us to neglect the future. One area of uncertainty concerns costs. The use of discounting as a principle for financing future projects is well established. Discounting is a means of making sufficient financial investment now to achieve sufficient finance for projects which, like a repository, will be developed at some future date. Discounting is based on the recognition that present money values depreciate so will be less in future. The calculation depends on two variables, the discount rate applied and the time period of the investment. The higher the discount rate the lower will the value be over time; likewise, the longer the time period the lower the value. Thus, for very long-term projects like a waste repository or decommissioning a power station, the future costs expressed in today's prices will be very low. By this means it can be argued the present generation is literally discounting the future. A major problem with long-term financing is uncertainty about the ultimate costs of a project which, in the case of a repository, may only be at a generic stage with no site in sight. In any case nuclear projects routinely have suffered cost overruns, delays and a tendency towards appraisal optimism. Given the unknowns and uncertainties involved in cost calculation, the idea of providing for a fixed unit cost for a project that is at present uncostable at a future date that is undetermined is both unreasonable and unjustifiable. The best that may be hoped for is that the costs of managing the legacy will be met through operators and governments making adequate provision against all eventualities and ensuring funds are made available to meet liabilities as and when they arise. There is an element of pious pleading that may be unrealistic it must be said. Whatever provisions are made it seems inevitable that a burden of cost, effort and risk will be laid on the future from the actions of previous and present generations.

Another area of uncertainty is timescale. Here it may be useful to distinguish between what may be called 'sociocultural timescales' which are relatively short and 'geoscientific timescales' which may stretch to infinity. Geoscientific time is the realm of scientists and engineers concerned with conceptualising and designing facilities such as a geological repository with containment systems that are demonstrably safe for immensely long time periods. Needless to say, at very long timescales the scientific problems of measuring and modelling outcomes become incommensurable. 'On sufficiently long time scales, any statement at all about the impacts of current actions and about obligations of current societies towards the future eventually become meaningless' (NEA/OECD, 2006, p.21). By contrast, sociocultural timescales are concerned with human perception and concern which extends over a short period, one or two generations, a hundred years or so at most (Duncan, 2002). In this timescale attention is more likely to be fixed on the immediate future, on ensuring safety through secure storage. Since the far future is difficult to envisage in any contemporary terms there is a tendency to accept that, to an extent, the future must be left to take care of itself. For example, in the UK, government is reasonably confident that coastal sites for new nuclear power stations and radioactive waste stores can be presumed safe from the impacts of sea level rise, storm surges and coastal processes arising from climate change for around a hundred years. Thereafter it is assumed that the radioactive wastes will have been removed for permanent burial in repositories which, given all the uncertainties about the longer term management, is a heroic assumption which short-changes the future.

The different perceptions of timescale lead to different interpretations of how the nuclear legacy should be managed to achieve intergenerational equity. It becomes a matter of when should we exercise our responsibility to the future? On one account we have a continuing responsibility which 'has to extend to the reach of the impact of our actions' (Adam, 2007, p.16). This view supports the concept of long-term storage which keeps the legacy visible and therefore enables remedial action if necessary. It leaves future generations free to take decisions on future management. But, it is also incumbent on the present generation, so far as is possible, to ensure the resources and information are available to enable future generations to make choices for waste management in the light of changing circumstances and knowledge. This approach places responsibility and freedom to decide on future generations. This continuing responsibility implying a strategy of 'leave it till later' requires a process of decision making in what may be called a 'continuing present' (Kommentus, 2001).

By contrast is the 'do it now' approach which takes the position that we should do what we can now to prevent passing burdens of risk, cost or effort on to the future. This approach favours removing the burden by placing the wastes in a deep disposal facility where they will be safely contained for an indefinite, effectively infinite time-span. This is a kind of precautionary approach though it is compromised by the problem of indeterminacy about the ultimate outcomes. The problem is 'we know virtually nothing about the living conditions of future generations – not to mention societal conditions hundreds, thousands, or even hundreds of thousands of years in the future' (KASAM, 2007, p.22). In effect this is an ethical approach tinged with pragmatism, in its recognition that we can only exert control over a short period and consequently the future in an ultimate sense must be left to itself. This approach of diminishing responsibility relies on decisions taken by the present generation and, so far as is possible, ensuring that information on what has been done can be passed on through the generations.

Ethical arguments can be addressed to either option of burying the legacy as soon as possible or continuing to store it. Continuing storage reflects a lack of confidence in disposal, the possibility of better options being developed in the future, a concern to keep the problem visible and flexibility in leaving the future with the opportunity to decide its own strategy. Disposal recognises the present should not bequeath burdens on the future generations who may be unwilling or incapable of dealing with the problem, and relies on the claim that wastes can be isolated from the accessible environment for the indefinite future. The concept of retrievability or phased disposal whereby wastes may be recovered from a disposal facility for a certain period of time was discussed in Chapter 4 (Aparicio, 2010). It appears as a compromise between immediate disposal and infinite storage apparently allowing more flexibility while removing or diminishing the burden of risk. But, given the long time spans, it offers, at most, a few hundred years. In any case, it is likely to be decades before deep disposal gets under way and over a hundred years before the existing legacy is tackled with the likelihood that decommissioning and associated waste arisings will still be scattered at various sites well into the next century and even beyond. And, if a new generation of reactors is built, the problem of managing wastes stretches into the far future, a time beyond human comprehension.

For the truth is that, whatever efforts are made to bury and forget the nuclear legacy, it will not go away but remain in sites in places like Hanford, Sellafield and La Hague for decades to come. So, for the foreseeable future, the next century, the long-term solution is already present: it is the safe and secure storage of the nuclear legacy that is already *in situ*. The priority remains clean-up, decontamination, remediation, vitrification, encapsulation, safe storage, the panoply of difficult, time-consuming and expensive processes that are being established as the nuclear industry has moved from production to clean-up. For the longer term, deep disposal remains an option, at present apparently the best but not the only option and, as time passes, other and better options may materialise. Given the timescales involved there is no need to hurry towards a disposal solution that may, in terms of proving a concept and finding a site, be difficult to implement. Society can, and should, take its time in dealing with its nuclear legacy. Meanwhile the focus should be on managing it where it is rather than a premature search for new places and possibly new communities for deep disposal. The problem we already have is difficult enough and will only be compounded if new reactors are built extending the timescales for implementation for very long, unknowable periods in the future. The burden of the existing legacy is unavoidable; we should not entertain having to deal with the avoidable wastes of a new build programme.

Managing the nuclear legacy is not just a technical problem; it is a social one, too. The communities I have studied over so many years, Hanford, Sellafield, La Hague and many others across the world, have long lived with the legacy and will continue to do so. In some places, Gorleben the most significant, the nuclear industry has met with resistance and has never become fully established. They are all, in their different ways, nuclear oases, peripheral places with distinctive identities. Their stories represent the changing discourses of the nuclear industry through its early years through to the present day. Whatever the future fortunes of the nuclear industry, its legacy and the communities that manage it will be with us for generations to come.

Note

1   Half-life is the time taken for half the radioactive nuclei in any sample to undergo radioactive decay.

References

Adam, B. (2007) *Ethics and Decision Making for Radioactive Waste,* Report for the Committee on Radioactive Waste Management (CoRWM), February.

Aparicio, L. (ed.) (2010) *Making Nuclear Waste Governable, Deep Underground Disposal and the Challenge of Reversibility,* Paris, Springer and Andra.

Blowers, A. (1999) 'Nuclear waste and landscapes of risk', *Landscape Research,* 24, 3, 241–264.

Brown, K. (2013) *Plutopia: Nuclear Families, Atomic Cities, and the Great Soviet and American Plutonium Disasters,* Oxford, Oxford University Press.

Camilleri, J. (1984) *The State and Nuclear Power,* Brighton, Wheatsheaf Books.

CoRWM (Committee on Radioactive Waste Management) (2005) *Deliberative Democracy and Decision Making for Radioactive Waste,* report for CoRWM, October.

CoRWM (2007) *Moving Forward: CoRWM's Proposals for Implementation,* CoRWM document 1703, February.

Crenson, M. (1971) *The Un-Politics of Air Pollution: A Study of Non-Decisionmaking in the Cities,* Baltimore, The Johns Hopkins Press.

Davies, H. (ed.) (2012) *Sellafield Stories: Life with Britain's First Nuclear Plant,* London, Constable.

Dryzek, J. (1997) *The Politics of the Earth,* Oxford, Oxford University Press.

Duncan, I. (2002) 'Disposal of radioactive waste: a puzzle in four dimensions', *Nuclear Energy,* 41, 1, February, 75–80.

Elliott, D. (ed.) (2007) *Nuclear or Not?* Basingstoke, Palgrave Macmillan.

Falk J. (1982) *Global Fission,* Oxford, Oxford University Press.

Goodin, R. (1992) *Green Political Theory,* Cambridge, Polity Press.

Goodin, R. (2003) *Reflective Democracy,* Oxford, Oxford University Press.

Habermas, J. (1981) *Theory of Communicative Action,* two volumes, vol. I translated 1984, vol. 2 1987 by Thomas McCarthy, Boston, Beacon Press.

Hecht, G. (1998) *The Radiance of France: Nuclear Power and National Identity after World War II,* Cambridge Mass. and London, The MIT Press.

International Atomic Energy Agency (1995) *The Principles of Radioactive Waste Management,* Safety Series No. 111-F, IAEA, Vienna.

Jonas, H. (1984) *The Imperative of Responsibility,* Chicago, University of Chicago Press.

KASAM (2007) *Nuclear Waste State-of-the-Art Report 2007 – Responsibility of Current Generation, Freedom of Future Generations,* main report from the Swedish National Council for Nuclear Waste (KASAM) Swedish Government Official Reports, Stockholm.

Kommentus (2001) *Responsibility, Equity and Credibility – Ethical Dilemmas Relating to Radioactive Waste,* Stockholm, Kommentus Forlag.

Loeb, P. (1986) *Nuclear Culture: Living and Working in the World's Largest Atomic Complex,* Philadelphia, New Society Publishers.

NEA/OECD (Nuclear Energy Agency/Organisation for Economic Cooperation and Development) (2006) *The Environmental and Ethical Basis of Geological Disposal,* Paris, NEA/OECD.

Pringle, P. and Spigelman, J. (1982) *The Nuclear Barons,* London, Sphere Books.

Rawles, K. (2007) In *Ethics and Decision Making for Radioactive Waste,* Report for the Committee on Radioactive Waste Management (CoRWM), February, pp.25–31, 62–69.

Renn, O. (2008) *Risk Governance: Coping with Uncertainty in a Complex World,* London, Earthscan.

Sanger, S. (1995) *Hanford and the Bomb: An Oral History of World War II,* Seattle, Living History Press.

Smith, G. (2003) *Deliberative Democracy and the Environment,* London, Routledge.

Sovacool, B. (2011) *Contesting the Future of Nuclear Power,* Singapore, World Scientific.

Tönnies, F. (1887) *Gemeinschaft und Gesellschaft,* Leipzig, Fues's Verlag, translated by C.P. Loomis (1997) as *Community and Society,* East Lansing, Michigan State University Press.

Weber, M. (1921) *Economy and Society,* ed. by Günter Roth and Claus Wittich, University of California Press (1921/1968/1978).

Wynne, B., Waterton, C. and Grove-White, R. (1993) *Public Perceptions and the Nuclear Industry in West Cumbria,* Centre for the Study of Environmental Change, Lancaster University.

Zonabend, F. (1993) *The Nuclear Peninsula,* Cambridge, Cambridge University Press.

INDEX

Taylor & Francis eBooks

Helping you to choose the right eBooks for your Library

Add Routledge titles to your library's digital collection today. Taylor and Francis ebooks contains over 50,000 titles in the Humanities, Social Sciences, Behavioural Sciences, Built Environment and Law.

Choose from a range of subject packages or create your own!

Benefits for you

>> Free MARC records
>> COUNTER-compliant usage statistics
>> Flexible purchase and pricing options
>> All titles DRM-free.

Benefits for your user

>> Off-site, anytime access via Athens or referring URL
>> Print or copy pages or chapters
>> Full content search
>> Bookmark, highlight and annotate text
>> Access to thousands of pages of quality research at the click of a button.

REQUEST YOUR **FREE** INSTITUTIONAL TRIAL TODAY

Free Trials Available
We offer free trials to qualifying academic, corporate and government customers.

eCollections – Choose from over 30 subject eCollections, including:

Archaeology	Language Learning
Architecture	Law
Asian Studies	Literature
Business & Management	Media & Communication
Classical Studies	Middle East Studies
Construction	Music
Creative & Media Arts	Philosophy
Criminology & Criminal Justice	Planning
Economics	Politics
Education	Psychology & Mental Health
Energy	Religion
Engineering	Security
English Language & Linguistics	Social Work
Environment & Sustainability	Sociology
Geography	Sport
Health Studies	Theatre & Performance
History	Tourism, Hospitality & Events

For more information, pricing enquiries or to order a free trial, please contact your local sales team:
www.tandfebooks.com/page/sales

Routledge
Taylor & Francis Group

The home of
Routledge books

www.tandfebooks.com